Photoacoustics and Photoacoustic Spectroscopy

CHEMICAL ANALYSIS

A SERIES OF MONOGRAPHS ON ANALYTICAL CHEMISTRY AND ITS APPLICATIONS

Editors

P. J. ELVING · J. D. WINEFORDNER

Editor Emeritus: I. M. KOLTHOFF

Advisory Board

F. W. Billmeyer	V. G. Mossotti
E. Grushka	A. L. Smith
B. L. Karger	B. Tremillon
V. Krivan	T. S. West

VOLUME 57

A WILEY-INTERSCIENCE PUBLICATION

JOHN WILEY & SONS

New York / Chichester / Brisbane / Toronto

Photoacoustics and Photoacoustic Spectroscopy

ALLAN ROSENCWAIG

University of California, Lawrence Livermore Laboratory

WITHDRAWN

BOWLING GREEN UNIVERSITY
APR 20 1981
LIBRARY
SERIALS DEPT.

A WILEY-INTERSCIENCE PUBLICATION

JOHN WILEY & SONS

New York / Chichester / Brisbane / Toronto

Bowling Green Univ. Library

Copyright © 1980 by John Wiley & Sons, Inc.

All rights reserved. Published simultaneously in Canada.

Reproduction or translation of any part of this work
beyond that permitted by Sections 107 or 108 of the
1976 United States Copyright Act without the permission
of the copyright owner is unlawful. Requests for
permission or further information should be addressed to
the Permissions Department, John Wiley & Sons, Inc.

Library of Congress Cataloging in Publication Data:

Rosencwaig, Allan, 1941-
    Photoacoustics and photoacoustic spectroscopy.

    (Chemical analysis; v. 54 ISSN 0069-2883)
    "A Wiley-Interscience publication."
    Includes bibliographical references and index.
    1. Optoacoustic spectroscopy.   I.   Title.
II. Series.
QD96.06R67   543'.0858   80-17286
ISBN 0-471-04495-4

Printed in the United States of America

10 9 8 7 6 5 4 3 2 1

PREFACE

We in the scientific community are accustomed to an almost daily announcement of new discoveries. While some of these turn out to be of fundamental and lasting significance, most simply become historical footnotes, illustrating the scientific fashion of their time. The photoacoustic effect was itself once an exciting new discovery, and then it, like so many other discoveries, was assigned to the scientific footnotes of the nineteenth century. However, almost 100 years later, the photoacoustic effect was "rediscovered," and it now appears to be well on its way to achieving lasting and significant status. In the 7 years since its "rediscovery" as a technique for the study of condensed and gaseous matter, photoacoustics already has established itself as an important research and analytical tool in many different areas, including physics, chemistry, biology, and medicine.

I have written this book to provide a reasonably comprehensive overview of both the theoretical and experimental aspects of this novel technique and of its many different applications. I have tried to make this book useful to both those new to the field and those already active in it. The first part of the book deals with the history of the photoacoustic effect and a review of both past and recent work in gaseous spectroscopy and vapor detection. For this latter topic I have, at times, relied on some of the excellent reviews in Yoh-Han Pao's book on *Optoacoustic Spectroscopy and Detection*. Most of this book, however, is devoted to photoacoustic studies, both spectroscopic and nonspectroscopic, on liquids and solids, for it is in this area that the field has been most active and fastest growing. There is, in the first part of the book, some new, previously unpublished material on the photoacoustic theory of condensed media. Here I have attempted to show how a simple thermoelastic theory can be used to extend the Rosencwaig-Gersho model and provide a more complete analysis of photoacoustic signal generation both in gas-microphone and in piezoelectric systems. I have tried to review, and place in perspective, most of the major publications on photoacoustics through 1978, and I apologize to those authors whose work I may have inadvertently omitted.

As a researcher active in photoacoustics since its "rediscovery," I have had the privilege of personally knowing many of the other investigators in the field. I would like to thank many of them for their help and advice. Of these I should like to mention particularly Eli Pines, John McClelland,

Alan McDonald, Grover Wetsel, and Allen Bard. I am also indebted to Mrs. M. McCaslin-Gregory for her unfailing assistance in the preparation of the manuscript. Finally, my deepest thanks to my wife Joan for her invaluable work, moral support, and wonderful patience.

ALLAN ROSENCWAIG

Danville, California
October 1980

CONTENTS

Photoacoustics and
Photoacoustic Spectroscopy

INTRODUCTION

In its broadest sense, spectroscopy can be defined as the study of the interaction of energy with matter. As such, it is a science encompassing many disciplines and many techniques. In the field of high-energy physics, the radiation is sufficiently energetic to seriously perturb, and in some cases, even transform the matter with which it interacts. On the other hand, in the oldest form of spectroscopy, optical spectroscopy, the energy is usually too low to perturb or noticeably alter the material under study. The energy used in optical spectroscopy exists in the form of optical photons or quanta, with wavelengths ranging from less than 1 Ångstrom in the x-ray region, to more than 100 microns (10^6 Å) in the far-infrared. Because of its versatility, range, and nondestructive nature, optical spectroscopy remains a widely used and most important tool for investigating and characterizing the properties of matter.

Conventional optical spectroscopies tend to fall into two major categories (Bosquet, 1971). The first category involves the study of the optical photons that are transmitted through the material of interest, that is, the study of those photons that did not interact with the material. The second category involves the study of the light that is scattered or reflected from the material, that is, those photons that have undergone some interaction with the material. Almost all conventional optical methods are variations on these two basic techniques. As such, they are distinguished not only by the fact that optical photons constitute the incident energy beam, but also by the fact that the data are obtained by detecting some of these photons after the beam has interacted with the matter or material under investigation. It should be noted that these optical techniques preclude the detection and analysis of those photons that have undergone an absorption, or annihilation, interaction with the material, even though this process is often the one of most interest to the investigator.

Optical spectroscopy has been a scientific tool for over a century, and it has proven invaluable in studies on reasonably clear media, such as solutions and crystals, and on specularly reflective surfaces. There are, however, several instances where conventional transmission spectroscopy is inadequate even for the case of clear, transparent materials. Such a situation arises when one is attempting to measure a very weak absorption, which in turn involves the measurement of a very small change in the intensity of a strong, essentially unattenuated, transmitted signal. Although this problem occurs for all forms of matter, it has received particular attention in the case of transparent gas mixtures containing minute quantities of an absorbing species or pollutant. Various techniques developed to overcome this difficulty, such as derivative spectroscopy, have proven to be generally inadequate.

In addition to weakly absorbing materials, there are a great many nongaseous substances, both organic and inorganic, that are not readily amenable to the conventional transmission or reflection modes of optical spectroscopy. These are usually highly light-scattering materials, such as powders, amorphous solids, gels, smears, and suspensions. Other difficult materials are those that are optically opaque and have dimensions that far exceed the penetration depth of the photons. In the former case, the optical signal is composed of a complex combination of specularly reflected, diffusely reflected, and transmitted photons, making the analysis of the data extremely difficult. In the latter case, the absorptive properties of the material are difficult, if not impossible, to determine, since essentially no photons are transmitted. Over the years, several techniques have been developed to permit optical investigation of highly light-scattering and opaque substances. The most common of these are diffuse reflectance (Wendlandt and Hecht, 1966), attenuated total reflection (ATR), and internal reflection spectroscopy (Wilks and Hirschfeld, 1968), and Raman scattering (Wright, 1969). All these techniques have proven to be very useful, yet each suffers from serious limitations. In particular, each method is applicable to only a relatively small category of materials, each is useful only over a small wavelength range, and the data obtained are often difficult to interpret.

During the past few years, another optical technique has been developed to study those materials that are unsuitable for the conventional transmission or reflection methodologies (Rosencwaig, 1973; 1975; 1978). This technique, called photoacoustic spectroscopy or PAS, is different than the conventional techniques chiefly in that even though the incident energy is in the form of optical photons, the interaction of these photons with the material under investigation is studied not through subsequent detection and analysis of some of the photons, but rather through a direct measure of the energy absorbed by the material as a result of its interaction with the photon beam.

Although more is said about experimental methodology later in this book, a brief description here of the photoacoustic technique seems appropriate. In photoacoustic spectroscopy, the sample to be studied is often placed in a closed cell or chamber. For the case of gases and liquids the sample generally fills the entire chamber. In the case of solids, the sample fills only a portion of the chamber, and the rest of the chamber is filled with a nonabsorbing gas such as air. In addition, the chamber also contains a sensitive microphone. The sample is illuminated with monochromatic light that either passes through an electromechanical chopper or is intensity modulated in some other fashion. If any of the incident photons are absorbed by the sample, internal energy levels within the sample are excited. Upon subsequent deexcitation of these energy levels, all or part of the absorbed photon energy is then transformed into heat energy through nonradiative deexcitation processes. In a gas this heat energy appears as kinetic energy of the gas molecules, while in a solid or liquid, it appears as vibrational energy of ions or atoms. Since the incident radiation is intensity modulated, the internal heating of the sample is also modulated.

Since photoacoustics measures the internal heating of the sample, it clearly is a form of calorimetry, as well as a form of optical spectroscopy. There are many calorimetric techniques by which one can detect and measure the heat produced during a physical or chemical process. The most obvious approach for the detection of heat production is to employ a conventional calorimeter based on the usual temperature sensors such as thermistors and thermopiles (Gelernt et al., 1974). These classical techniques, though simple and well developed, have several inherent disadvantages for photoacoustic spectroscopy in terms of sensitivity, detector rise time, and the speed at which measurements can be made. More suitable calorimetric techniques measure heat production through volume and pressure changes produced in the sample or in an appropriate transducing material in contact with the sample.

In gaseous samples, volume changes can be quite large as a result of internal heating. In these cases, a displacement-sensitive detector such as a capacitor microphone proves to be an excellent heat detector. With present microphones and associated electronics it is possible to detect temperature rises in a gas of 10^{-6}°C, or a thermal input of the order of 10^{-9} calories/cm^3-sec. The primary disadvantage with a detector that responds to volume changes is that the response time is limited both by the transit time for a sound wave in the gas within the cell cavity and by the relatively low-frequency response of the microphone. Together, these two factors tend to limit the response time of a gas-microphone system to the order of 100 μsec or longer.

When dealing with liquids, or bulk solid samples, it is possible to measure heat production through subsequent pressure or stress variations in the sample itself by means of a piezoelectric detector in intimate contact with the sample. With these detectors, temperature changes of 10^{-7} to 10^{-6}°C can be detected, which for typical solids or liquids corresponds to thermal inputs of the order of 10^{-6} calories/cm³-sec. It should be borne in mind that because the coefficient of volume expansion of liquids and solids is 10–100 times smaller than that of gases, measurement of the heat production in liquids and solids directly with a displacement-sensitive detector such as a microphone would be 10–100 times less sensitive than using a pressure-sensitive device such as a piezoelectric detector.

It is, of course, not always possible to employ a piezoelectric detector, as in the case of a powdered sample, or a smear or gel. In these cases, a gas is used as a transducing medium, coupling the sample to a microphone. The periodic heating of the sample from the absorption of the optical radiation results in a periodic heat flow from the sample to the gas, which itself does not absorb the optical radiation. This in turn produces pressure and volume changes in the gas that drive the microphone. This method is not as direct as a contact piezoelectric measurement, but it is quite sensitive for solids with large surface/volume ratios, such as powders, and is capable of detecting temperature rises of 10^{-6} to 10^{-5}°C in such samples, or thermal inputs of about 10^{-5} to 10^{-6} calories/cm³-sec.

It is clear from the discussion above that photoacoustics is a combination of optical spectroscopy and calorimetry. Although a more appropriate name for the technique might be photocalorimetry, we continue to use the term photoacoustics for this methodology whether it employs microphones or piezoelectric detectors.

There are several advantages to photoacoustics as a form of spectroscopy, that is, when it is used to perform photoacoustic spectroscopy. Since *absorption* of optical or electromagnetic radiation is required before a photoacoustic signal can be generated, light that is transmitted or elastically scattered by the sample is not detected and hence does not interfere with the inherently absorptive PAS measurements. This is of crucial importance when one is working with essentially transparent media, such as pollutant-containing gases, that have few absorbing centers. The insensitivity to scattered radiation also permits the investigator to obtain optical absorption data on highly light-scattering materials, such as powders, amorphous solids, gels, and colloids. Another advantage is the capability of obtaining optical absorption spectra on materials that are completely opaque to transmitted light since the technique does not depend on the detection of photons. Coupled with this is the capability, unique to photoacoustic spectroscopy, of performing nondestructive depth-profile analysis of absorption as a function of depth into a material.

Furthermore, since the sample itself constitutes the electromagnetic radiation detector, no photoelectric device is necessary, and thus studies over a wide range of optical and electromagnetic wavelengths are possible without the need to change detector systems. The only limitations are that the source be sufficiently energetic (at least $10 \ \mu W/cm^2$) and that whatever windows are used in the system be reasonably transparent to the radiation. Finally, the photoacoustic effect results from a radiationless energy-conversion process and is therefore complementary to radiative and photochemical processes. Thus PAS itself may be used as a sensitive, though indirect, method for studying the phenomena of fluorescence and photosensitivity in matter.

Photoacoustics is, however, much more than spectroscopy. It is, after all, a photocalorimetric method that measures how much of the electromagnetic radiation absorbed by a sample is actually converted to heat. As such, photoacoustics can be used to measure the absorption or excitation spectrum, the lifetime of excited studies, and the energy yield of radiative processes. These are all spectroscopic measurements. On the other hand photoacoustics can also be used to measure thermal and elastic properties of materials, to study chemical reactions, to measure thicknesses of layers and thin films, and to perform a variety of other nonspectroscopic investigations. In such studies, the calorimetric or acoustic aspect of photoacoustics plays the dominant role, while the photon or electromagnetic part is simply a convenient excitation mechanism.

With its various spectroscopic and nonspectroscopic attributes, photoacoustics has already found many important applications in the research and characterization of materials. Photoacoustic studies are performed on all types of materials, inorganic, organic, and biological, and on all three matter states—gas, liquid, and solid. We explore many of the methodologies and applications of this new and exciting science in this book.

REFERENCES

Bosquet, P. (1971). *Spectroscopy and Its Instrumentation*, Crane-Russack, New York.

Gelernt, B., Findeisen, A., and Poole, J. J. (1974). *J. Chem. Soc., Faraday Trans. II*, **70**, 939.

Rosencwaig, A. (1973). *Opt. Commun.* **7**, 305.

Rosencwaig, A. (1975). *Anal. Chem.*, **47**, 592A.

Rosencwaig, A. (1978). In *Advances in Electronics and Electron Physics*, Vol. 46 (L. Marton, Ed.), pp. 207–311, Academic Press, New York.

Wendlandt, W. W., and Hecht, H. G. (1966). *Reflectance Spectroscopy*, Wiley, New York.

Wilks, P. A., Jr., and Hirschfeld, T. (1968). *Appl. Spectrosc. Rev.* **1**, 99.

Wright, G. B. (Ed.) (1969). *Light Scattering of Solids*, Springer-Verlag, Berlin and New York.

2

HISTORY OF PHOTOACOUSTICS

2.1 PREHISTORY—NINETEENTH CENTURY

Photoacoustic spectroscopy made its official debut in 1973. However, the concept on which it is based is quite old and an analogous technique, commonly referred to as optoacoustic spectroscopy, has been used for many years in the study of optical absorption phenomena in gases. The change of name from optoacoustic to photoacoustic was instituted by this author to reduce confusion with the acoustooptic affect in which light interacts with acoustic or elastic waves in a crystal. The term photoacoustic is now accepted as applying to those investigations that deal with nongaseous matter, while the term optoacoustic is still used in many circles for gas studies. For the sake of simplicity and clarity, we use the term photoacoustic for all the photocalorimetric studies discussed in this book no matter what the sample might be.

The photoacoustic effect in both nongaseous and gaseous matter was discovered in the nineteenth century and was first reported in 1880, when Alexander Graham Bell gave an account to the American Association for the Advancement of Science of his work on the photophone (Bell, 1880). In his paper, he briefly reported the accidental discovery of the photoacoustic effect in solids.

In brief, Bell's photophone consisted of a voice-activated mirror, a selenium cell, and an electrical telephone receiver. A beam of sunlight was intensity modulated at a particular point by means of the voice-activated mirror, as depicted in Figure 2.1. This intensity-modulated beam was then focused onto a selenium cell as shown in Figure 2.2. The selenium cell was itself incorporated in a conventional electrical telephone circuit. Since the electrical resistance of selenium varies with the intensity of light falling on it, the voice-modulated beam of sunlight resulted in electrically reproduced telephonic speech.

While experimenting with the photophone, Bell discovered that it was possible to obtain, at times, an audible signal directly, that is, in a nonelectrical fashion. This phenomenon occurred if the beam of sunlight was rapidly interrupted, say at 1000 Hz, as with a rotating slotted disk, and

7

Figure 2.1 The transmitter of Bell's photophone. A beam of sunlight is reflected from a voice-activated mirror through an aperture, thus imposing voice modulation on the beam.

then focused on solid substances such as the selenium, or onto solids that were in the form of diaphragms connected to a hearing tube.

In a publication in 1881 (Bell, 1881), Bell described in considerable detail his further investigations of this new effect. He found, for example, that if solid matter were placed inside a closed glass tube to which a hearing tube was attached, a quite audible signal could be detected if the material in the tube were illuminated with a rapidly interrupted beam of sunlight. He noted that "the loudest signals are produced from substances in a loose, porous, spongy condition, and from those that have the darkest or most absorbent colours" (Bell, 1881, p. 515).

In a series of definitive experiments (Bell, 1881), Bell demonstrated that the photoacoustic effect in solids was dependent on the absorption of light, and that the strength of the acoustic signal was in turn dependent on how strongly the incident light was absorbed by the material in the cell. He concluded that "the nature of the rays that produce sonorous effects in different substances depends upon the nature of the substances that are

Figure 2.2 The receiver of Bell's photophone. The voice-modulated sunlight beam is focused on the selenium cell whose changing resistance electrically modulates the telephonic signal.

exposed to the beam, and that the sounds are in every case due to those rays of the spectrum that are abosrbed by the body" (Bell, 1881). Bell thus correctly deduced the intrinsic optical *absorptive* dependence of the photo-acoustic effect.

In addition to studying the photoacoustic effect with solids, Bell and his associate, Sumner Tainter, also investigated the effect in liquids and gases (Bell, 1881). They observed that only weak signals were produced when the cell was filled with a light-absorbing liquid, but that quite strong signals were obtained when the cell was filled with light-absorbing gases. This difference in signal strength is quite understandable in light of our comments in Chapter 1. In Bell's experiments, the detector was a displacement-sensitive device, the ear. This was a highly efficient way for detecting the photoacoustic signal with gas samples. However, as we stated in Chapter 1, it is not as efficient a detector for liquid samples in spite of the fact that air is used as a coupling medium between the sample and the ear. Photo-acoustic experiments on gases were also performed by John Tyndall

(Tyndall, 1881) and Wilhelm Roentgen (Roentgen, 1881), who had heard of Bell's discovery of the previous year. They found, as had Bell, that the photoacoustic effect in light-absorbing gases could be readily observed for many different gaseous substances.

The series of papers by Bell, Tyndall, and Roentgen described experiments in which optical radiation, from the sun or a mercury arc, was chopped or intensity modulated by passing it through a rotating slotted disk and then directed into a closed chamber containing a sample. The sample was sometimes a solid or liquid, but most often it was a gas that was either colored, and thereby could absorb visible radiation, or colorless, but infrared absorbing. The signal produced in the enclosed chamber was detected by the ear as audible sound through a hearing tube connected to the chamber. Clearly then, the absorption of light by the sample resulted in pressure fluctuations of the air in the hearing tube, fluctuations having the same frequency as the chopping or modulating frequency.

The situation for gaseous samples was fairly well understood in the 1880s, since the basic gas laws were already well known. It was correctly assumed that the gaseous sample absorbed all or part of the chopped incident radiation and, by so doing, was itself periodically heated. Since the sample chamber was sealed off by a compliant diaphragm in the hearing tube, the periodic heating of the gas in the chamber resulted in both pressure and volume changes. These in turn were transmitted by the diaphragm of the hearing tube to its own air column and thence to the ear. A more complete, modern theory of the photoacoustic effect in gases is presented in Chapter 3.

Few experiments were reported on liquids, and no attempt was made by the nineteenth century investigators to explain the production of a photoacoustic signal in liquids. On the other hand, several attempts were made to account for the phenomenon in solids, although only recently has a satisfactory quantitative theory been formulated.

In attempting to account for the audible signal obtained from his experiments with dark spongy solids, such as lampblack, Bell hypothesized that:

"When a beam of sunlight falls upon the mass, the particles of lampblack are heated, and consequently expand, causing a contraction of the air-spaces or pores among them. Under these circumstances a pulse of air should be expelled, just as we would squeeze water out of a sponge. The force with which the air is expelled must be greatly increased by the expansion of the air itself, due to contact with the heated particles of lampblack. When the light is cut off, the converse process takes place. The lampblack particles cool and contract, thus enlarging the air-spaces among them, and the enclosed air also becomes cool. Under these circumstances, a partial vacuum should be formed among the particles, and the outside air would then be absorbed, as water is by a sponge when the pressure of the hand is removed. I imagine that in some such manner as this a wave of condensation is started in the atmosphere each time a beam of sunlight falls upon the lampblack, and a wave of rarefraction is originated when the light is cut off" (Bell, 1881, pp. 515–516).

In the case when the illuminated solid sample was in the form of a thin flexible membrane or disk, Bell supported the theory of Lord Rayleigh (Rayleigh, 1881), who concluded that the primary source for the photoacoustic signal was the mechanical vibration of the disk resulting from uneven heating of the disk when it was struck by the beam of sunlight.

Bell's conjectures were only partially correct. For example, we have found from experiments in which the photoacoustic cell was first thoroughly evacuated and then refilled with nonabsorbing noble gases, and from experiments with two-dimensional solids and other materials with weak adsorption properties, that adsorbed gases usually do not play a significant role in the production of the photoacoustic signal. Furthermore, it can be readily shown that thermal expansion and contraction of the solid sample, and in fact any optothermally induced mechanical vibrations of the solid, are generally too small in magnitude to be the sole cause of the usually observed acoustic signals.

In the opinion of the present author the hypotheses of Mercadier (1881) and Preece (1881) came closest to the modern explanation. Mercadier, who also experimented with the photoacoustic effect, suggested that the sound is due to "vibrating movement determined by the alternate heating and cooling produced by the intermittent radiations, principally in the gaseous layer adhering to the solid surface hit by these radiations" (Mercadier, 1881, p. 410). Preece wrote that the photoacoustic effect "is purely an effect of radiant heat, and it is essentially one due to the changes of volume in vapours or gases produced by the degradation and absorption of this heat in a confined space" (Preece, 1881, p. 517). He further noted that "Cigars, chips of wood, smoke, or any absorbent surfaces placed inside a closed transparent vessel will, by first absorbing and then radiating heat rays to the confined gas, emit sonourous vibrations" (Preece, 1881, p. 518).

From both experimental and theoretical considerations, we have concluded that the primary source of the acoustic signal obtained with a nongaseous sample in a photoacoustic cell that employs a microphone and a coupling gaseous medium arises from the periodic heat flow from the sample to the surrounding gas as the sample is cyclically heated by the incident light. We are thus in basic agreement with the thinking of Mercadier and Preece some 90 years ago.

After the initial flurry of interest generated by Alexander Graham Bell's original work, experimentation with the photoacoustic effect apparently ceased. The effect was obviously considered as being no more than an interesting curiosity of no great scientific or practical value. Furthermore, the experiments were difficult to perform and quantitate since they required the investigator's ear to be the signal detector.

2.2 "MODERN" HISTORY—TWENTIETH CENTURY

The photoacoustic effect lay completely dormant for nearly 50 years, until the advent of the microphone. Then in 1938 Viengerov (1938), working at the State Optical Institute, Leningrad, began using the phenomenon to study infrared light absorption in gases and to evaluate concentrations of gaseous species in gas mixtures. His light sources were blackbody infrared sources, such as Nernst glowers, and he employed an electrostatic microphonic arrangement whereby he measured the voltage change between charged capacitive microphone diaphragms. Viengerov was able to measure CO_2 concentrations in N_2 down to ~ 0.2 vol % using this photoacoustic method. Measurements of lower concentrations were limited both by the relatively low sensitivity of his microphone and by background absorption of the incident radiation by the cell windows and walls. That is, his photoacoustic measurements on the enclosed gaseous sample were seriously perturbed by the presence of an unwanted photoacoustic effect in the solid windows. As we see in Chapter 5, this is a problem that still plagues researchers performing photoacoustic spectroscopy of gaseous substances.

A year later, Pfund (1939) described a gas analyzer system in use at Johns Hopkins Hospitals, Baltimore, for measuring concentrations of CO and CO_2. Pfund's experiments are of additional interest because, instead of detecting pressure–volume changes with a microphone, he measured the corresponding changes in the gas temperature directly, using a thermopile shielded from the direct optical radiation. This instrument had a sensitivity comparable to Viengerov's initial apparatus.

A major improvement in the sensitivity of gas concentration analysis occurred in 1943 when Luft described a commercial automatically recording gas analyzer that employed two photoacoustic cells in a differential design. One cell contained the gas mixture to be analyzed, while the other contained the gas mixture minus the particular species of interest. In this instrument, therefore, the microphone output was proportional to the pressure difference between the two cells. Luft's differential analyzer had two major improvements over Viengerov's original design. First, it minimized the signal due to background absorption in the cell windows and walls, since the same background signal was present in both cells. Second, it permitted analysis of gas mixtures containing more than two species. Gas analyzers based on Luft's design became commercially available in 1946. These instruments had a sensitivity that permitted the measurement of CO_2 in N_2 down to a few parts per million (ppm), as compared to Viengerov's early capability of only a few parts per thousand (ppt).

It is most intriguing that the strong rebirth of the photoacoustic effect in the years following 1938 was apparently entirely limited to gases. It was not until the early 1970s, some 90 years after Bell's original discovery, that the photoacoustic effect in nongaseous matter was "rediscovered."

However, to return to our pre-1970 "modern" history, gas photoacoustic investigations included spectroscopy, as well as constituent analysis. A variation of Luft's design for a gas analyzer was also used to construct an infrared gas spectrometer, called the spectrophone (Viengerov, 1945). This instrument utilized infrared light sources, such as Nernst glowers, an infrared dispersive monochrometer to obtain monochromatic radiation, and a differential photoacoustic cell to obtain infrared absorption spectra of gases and vapors.

In addition to straightforward infrared absorption spectroscopy, scientists in the 1940s also used the photoacoustic effect to study deexcitation and energy-transfer processes in gases. They recognized that the photoacoustic signal is essentially a calorimetric measure of the amount of radiant energy absorbed by the gaseous sample that is dissipated through nonradiative or heat-producing processes. As such, the photoacoustic effect can be used to study this ubiquitous channel of energy-level deexcitation. Gorelik (1946) first proposed that measurement of the phase of the photoacoustic signal could be used to investigate the rate of energy transfer between the vibrational and the translational degrees of freedom of gas molecules. Such an energy transfer occurs through interatomic collisions, and thus the energy of the excited optical state eventually becomes translational or kinetic energy of the gas molecules. Gorelik's proposal was successfully put into practice by Slobodskaya (1948), and the use of the photoacoustic effect to study vibrational lifetimes of gaseous molecules quickly became an established technique. Further details of this technique are presented in Chapters 3 and 5.

Between 1950 and 1970 the photoacoustic gas analyzer employing a conventional light source gave way to the more sensitive gas chromatograph. And the spectrophone was overtaken by the more versatile infrared spectrophotometer. During this period, the photoacoustic effect was primarily employed to study vibrational lifetimes and other aspects of radiationless deexcitation in gases. In later chapters, we show that the advent of the laser provided a major impetus to photoacoustic spectroscopy in the early 1970s, and once again photoacoustic gas analyzers and spectrometers found exciting uses.

It is, as we mention earlier, most intriguing that this rebirth of photoacoustics was apparently limited to gases only. How, in spite of the many years of work with gases since 1938, solid- and liquid-state photoacoustic studies remained forgotten until the 1970s remains a mystery. In the second half of this book we detail the strong reemergence of nongaseous photoacoustics that occurred after 1973 and illustrate its development into the powerful technique of today.

REFERENCES

Bell, A. G. (1880). *Am. J. Sci.* **20**, 305.

Bell A. G. (1881). *Philos. Mag.* **11**, (5) 510.

Gorelik, G. (1946). *Dokl. Akad. Nauk SSSR* **54**, 779.

Luft, K. F. (1943). *Z. Tech. Phys.* **24**, 97.

Mercadier, M. E. (1881). *C. R. Hebd. Serv. Acad. Sci.* **92**, 409.

Pfund, A. H. (1939). *Science* **90**, 326.

Preece, W. H. (1881). *Proc. R. Soc. Lond.* **31**, 506.

Rayleigh (Lord) (1881). *Nature (Lond.)* **23**, 274.

Roentgen, W. C. (1881). *Philos. Mag.* **11**, (5) 308.

Slobodskaya, P. V. (1948). *Izv. Akad. Nauk SSSR, Ser. Fiz.* **12**, 656.

Tyndall, J. (1881). *Proc. R. Soc. Lond.* **31**, 307.

Viengerov, M. L. (1938). *Dokl. Akad. Nauk SSSR* **19**, 687.

Viengerov, M. L. (1945). *Dokl. Akad. Nauk SSSR* **46**, 182.

THEORY OF PAS OF GASES

3.1 INTRODUCTION

During the nineteenth century, investigators of the photoacoustic effect concentrated their studies on gaseous samples. Experiments with gases were the easiest to perform and were also understood. Even for most of the twentieth century, the photoacoustic technique has been applied exclusively to gaseous substances.

In modern photoacoustic spectroscopy of gases, intensity-modulated light either from a laser or a more conventional source enters a photoacoustic cell filled with the gaseous sample as shown in Figure 3.1. All or a portion of the incident radiation is absorbed by the gas resulting in a pressure–sound disturbance that is converted into an electrical signal by means of a microphone. The steps in the generation of the photoacoustic signal are depicted in Figure 3.2.

3.2 ABSORPTION OF LIGHT

When a gas molecule absorbs a photon, it goes from its ground state E_0 to an excited state E_1, the energy difference between the states being $E_1 - E_0 = h\nu$, where ν is the frequency of the absorbed photon. The molecule can then lose this energy and return to the ground state in the following ways:

1. It can reradiate a photon—radiative deexcitation.
2. It can initiate a photochemical event such as bond rearrangement—photochemistry.
3. It can collide with another molecule of the same species that is in the ground state E_0 and excite it to its excited state E_1—intersystem energy transfer.
4. It can collide with any other molecule in the gas and transfer energy to translational or kinetic energy shared by both molecules—heating.

The sound wave detected by the microphone results from the fourth process, since an increased kinetic energy of the gas molecules is simply

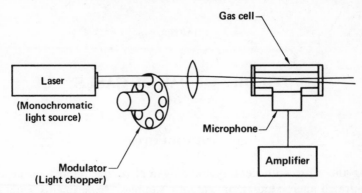

Figure 3.1 Block diagram of a modern gas photoacoustic spectrometer. (Reproduced by permission, Kreuzer, L. B., 1977.)

increased heat energy, or an increased temperature of the gas. If the incident photon radiation is intensity modulated at a rate that is slow compared to the rate of process 4 above, then the optical modulation results in a coherent modulation in the temperature of the gaseous sample. From the gas laws it is clear that in the enclosed volume of the photoacoustic cell, the modulation of the gas temperature results in a periodic pressure fluctuation with a modulation frequency equal to the optical modulation frequency. This pressure fluctuation is, of course, a sound wave and as such is readily detectable with a microphone.

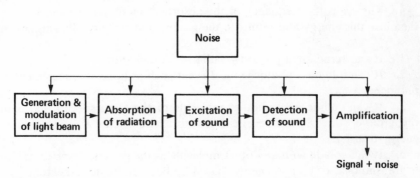

Figure 3.2 The steps in the generation of a photoacoustic signal. (Reproduced by permission, Kreuzer, L. B., 1977.)

3.2.1 Rate Equation for a Two-Level System

Let us now consider a simple mathematical formulation used to describe the optical absorption process in a gas. For simplicity we consider here a two-level system such as that depicted in Figure 3.3.

The parameter r_{ij} represents a radiative transition rate from level i to level j, while c_{ij} represents a nonradiative collisionally induced transition from E_i to E_j. Clearly,

$$r_{ij} = \rho_\nu B_{ij} + A_{ij} \qquad (3.1)$$

where ρ_ν is the radiation density at energy $E_\nu = h\nu = E_1 - E_0$, where E_1 and E_0 are the energies of the excited and ground states, respectively; $B_{ij} = B_{ji}$ and represents the Einstein coefficients for stimulated emission $i \rightarrow j$; and A_{ij} represents the Einstein coefficients for spontaneous emission $i \rightarrow j$, and thus in our model $A_{01} = 0$ since $E_1 > E_0$. Let us now consider the time dependence of N_1, the number of molecules per unit volume in energy state E_1. The rate of change of N_1 with time, $\dot{N}_1$, is simply the number of molecules entering level 1 minus the number leaving level 1 per unit time. Thus

$$\dot{N}_1 = (r_{01} + c_{01})N_0 - (r_{10} + c_{10})N_1$$

$$= (\rho_\nu B_{01} + A_{01} + c_{01})N_0 - (\rho_\nu B_{10} + A_{10} + c_{10})N_1 \qquad (3.2)$$

But $A_{01} = 0$ and $c_{01} \simeq 0$ since there is a very low probability of a collisional

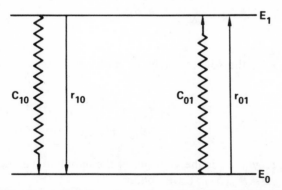

Figure 3.3 Schematic representation of a two-level electronic system showing the radiative and nonradiative transitions. (Reproduced by permission, Stettler, J. D. and Witriol, N. M., 1977.)

excitation of an atom from level 0 to level 1 at room temperature. In addition $B_{01} = B_{10}$. Thus

$$\dot{N}_1 = \rho_\nu B_{10}(N_0 - N_1) - (A_{10} + c_{10})N_1$$
$$= \rho_\nu B_{10}(N_0 - N_1) - (\tau_r^{-1} + \tau_c^{-1})N_1 \qquad (3.3)$$

where $\tau_r = 1/A_{10}$ is the relaxation time for a radiative transition from level 1, and $\tau_c = 1/c_{10}$ is the relaxation time for a nonradiative or collisional transition from level 1.

Defining a total relaxation time by τ where

$$\tau^{-1} = \tau_r^{-1} + \tau_c^{-1} \qquad (3.4)$$

we obtain

$$\dot{N}_1 = \rho_\nu B_{10}(N_0 - N_1) - \tau^{-1} N_1 \qquad (3.5)$$

Similarly, for N_0

$$\dot{N}_0 = -\rho_\nu B_{10}(N_0 - N_1) + \tau^{-1} N_1 \qquad (3.6)$$

Thus

$$\frac{d}{dt}(N_1 - N_0) = -2\rho_\nu B_{10}(N_1 - N_0) - 2\tau^{-1} N_1 \qquad (3.7)$$

At steady state $d/dt(N_1 - N_0) = 0$. Thus

$$N_1 = \frac{\rho_\nu B_{10} N}{2\rho_\nu B_{10} + \tau^{-1}} \qquad (3.8)$$

where $N = N_1 + N_0$ the total number of molecules per cubic centimeter. Similarly,

$$N_0 = \frac{(\rho_\nu B_{10} + \tau^{-1})N}{2\rho_\nu B_{10} + \tau^{-1}} \qquad (3.9)$$

Now $\rho_\nu = Ih\nu/c$, where I is the intensity of the light and c is the velocity of light. Defining $B = B_{10}(h\nu/c)$, we obtain

$$N_1 = \frac{BIN}{2BI + \tau^{-1}} \quad \text{and} \quad N_0 = \frac{(BI + \tau^{-1})N}{2BI + \tau^{-1}} \qquad (3.10)$$

In the case where the incident radiation is modulated at a frequency ω we can set

$$I = I_0(1 + \delta e^{i\omega t}) \qquad 0 \leqslant \delta \leqslant 1 \tag{3.11}$$

This then gives

$$N_1 = N \frac{BI_0(1 + \delta e^{i\omega t})}{2BI_0(1 + \delta e^{i\omega t}) + \tau^{-1}} \tag{3.12}$$

3.3 EXCITATION OF ACOUSTIC WAVE

The incident radiation I, given by (3.11), pumps a significant number of molecules from the ground state E_0 to the excited state E_1. When the excited molecules decay to the ground state by means of collision (the c_{ij} terms), the energy difference $\Delta E = E_1 - E_0$ goes into translational energy, that is, the velocities of the colliding molecules increase.

Ignoring the rotational and vibrational energies, the total internal energy of the gas per unit volume is given by

$$U = \sum_i (N_i E_i) + K \tag{3.13}$$

where the sum is over all levels i, and K is the translational or kinetic energy per cubic centimeter. In our simple two-level model

$$U = N_1 E_1 + K \tag{3.14}$$

and

$$\dot{U} = \dot{N}_1 E_1 + \dot{K} \tag{3.15}$$

Now by conservation of energy, $\dot{U}$ is simply equal to the difference between the energy absorbed and the energy reradiated. Thus

$$\dot{U} = (r_{01} N_0 - r_{10} N_1) E_1 \tag{3.16}$$

and

$$\dot{K} = c_{10} N_1 E_1 \tag{3.17}$$

From thermodynamics we have

$$dK = \left(\frac{\partial K}{\partial T}\right)_V dT + \left(\frac{\partial K}{\partial V}\right)_T dV \tag{3.18}$$

where T and V are the temperature and volume, respectively. For constant volume

$$dK = \left(\frac{\partial K}{\partial T}\right)_V dT = C_V dT \tag{3.19}$$

where C_V is the specific heat per unit volume at constant volume. Therefore,

$$K = C_V T + f(V) \tag{3.20}$$

where $f(V)$ is some function that is dependent on volume but not on temperature. For an ideal gas, the pressure is given by

$$p = NkT \tag{3.21}$$

where k is the Boltzmann constant. Thus

$$p = \frac{k}{C_v} N[K - f(V)] \tag{3.22}$$

The pressure wave is given by $\partial p/\partial t$. Taking the derivative of (3.22).

$$\dot{p} = \frac{k}{C_v} N\dot{K} = \frac{k}{C_v} N(c_{10} N_1 E_1) \tag{3.23}$$

From (3.12) we then have

$$\dot{p} = \frac{k}{C_v} \frac{N^2 E_1}{\tau_c} \frac{BI_0(1 + \delta e^{i\omega t})}{2BI_0(1 + \delta e^{i\omega t}) + \tau^{-1}} \tag{3.24}$$

Expanding in powers of $(\delta e^{i\omega t})$ and retaining only the $e^{i\omega t}$ term, we obtain

$$\dot{p} = \frac{kE_1 N^2}{C_v} \left\{ \frac{2\tau_c^{-2} BI_0 \delta}{(2BI_0 + \tau^{-1})\left[(2BI_0 + \tau^{-1})^2 + \omega^2\right]^{1/2}} \right\} e^{i(\omega t - \gamma)} \tag{3.25}$$

where $\gamma = \omega\tau$. Integrating, we then obtain

$$p = \frac{kE_1N^2}{C_v\omega}\left\{\frac{2\tau_c^{-2}BI_0\delta}{(2BI_0+\tau^{-1})\left[(2BI_0+\tau^{-1})^2+\omega^2\right]^{1/2}}\right\}e^{i(\omega t-\gamma-\pi/2)} \quad (3.26)$$

The photoacoustic microphone signal measured is

$$q = -p \quad (3.27)$$

It is of interest to consider two limiting light-intensity cases. In the first and most usual case, I_0 is small and the optical pumping term

$$2BI_0 \ll \tau^{-1}$$

Then

$$q \simeq \frac{kE_1N^2}{C_v\omega}\left(\frac{\tau}{\tau_c}\right)^2\frac{2BI_0\delta}{(1+\omega^2\tau^2)^{1/2}}e^{i(\omega t-\gamma+\pi/2)} \quad (3.28)$$

The photoacoustic signal is thus proportional to N^2, or the density of the gas squared, and varies linearly with the light intensity I_0. It is also proportional to the term $(\tau/\tau_c)^2$, where τ is the total deexcitation lifetime and τ_c is the collisional lifetime. Since τ_c decreases with increasing temperature the photoacoustic signal generally increases with increasing temperature.

For high optical intensities of the order of W/cm^2, $2BI_0 \gg \tau^{-1}$, and the photoacoustic signal becomes for small ω

$$q \simeq \frac{kE_1N^2}{C_v\omega}\tau_c^{-2}\frac{1}{BI_0}\delta e^{i(\omega t-\gamma+\pi/2)} \quad (3.29)$$

Here absorption saturation occurs where the signal varies as I_0^{-1}. This saturation is simply the result of attempting to pump harder than the upper level can be deexicited $(BI_0 > \frac{1}{2}\tau^{-1})$. Experimental evidence on the effects of absorption saturation on the photoacoustic signal has been obtained by Pao and Claspy (1975).

3.4 ENEGY TRANSFER PHYSICS

In a single kinetic collision between two gas molecules, one can compute the probability p_{ij} that an energy transfer occurs from state E_i to state E_j.

The relation between the collisional deexcitation rate c_{ij} and p_{ij} is given by

$$c_{ij} = \overline{Z}_{AB} p_{ij} \qquad (3.30)$$

where $\overline{Z}_{AB}$ is the average number of gas kinetic collisions per cubic centimeter per second between two molecular species A and B. The quantity $\overline{Z}_{AB}$ is calculated on the assumption that only binary collisions are important and that energy level populations follow a Boltzmann distribution (Taylor and Bitterman, 1969; Herzfeld and Litovitz, 1959). Using these assumptions we find that for a vibrational→translational energy transfer

$$\overline{Z}_{AB} = \left(\frac{8\pi kT}{\mu_{AB}} \right)^{1/2} \sigma_{AB}{}^2 (1 - e^{-h\nu/kT}) \qquad (3.31)$$

where $h\nu = E_i - E_j$, μ_{AB} is the reduced mass given by

$$\mu_{AB} = \frac{m_A m_B}{m_A + m_B} \qquad (3.32)$$

and $\sigma_{AB}{}^2$ is the collision cross section.

$$\sigma_{AB} = \sigma_A + \sigma_B \qquad (3.33)$$

where σ_A and σ_B are the repulsive radii of molecules A and B, respectively.

Similarly, for a vibrational→vibrational energy transfer,

$$\overline{Z}_{AB} = \left(\frac{8\pi kT}{\mu_{AB}} \right)^{1/2} \sigma_{AB}{}^2 \qquad (3.34)$$

Energy-transfer rates p_{ij} depend on the amount of energy transferred, as well as on the transfer process. Electronic energies are of the order of

$$E_{el} \simeq \frac{\hbar}{ma^2} \qquad (3.35)$$

where m and a are the mass and orbital radius of the optically excited electron, respectively. These energies are in the ultraviolet and visible regions of the optical spectrum.

Molecular vibrational energies are of the order of

$$E_{vib} \simeq \left(\frac{m}{M} \right)^{1/2} E_{el} \qquad (3.36)$$

where m is the electronic mass and M is the molecular mass. Since $(m/M) \simeq 10^{-4}$ for most molecules, vibrational energies are in the infrared region of the spectrum. Rotational energies are even smaller and are of the order of

$$E_{\text{rot}} \simeq \left(\frac{m}{M}\right) E_{\text{el}} \qquad (3.37)$$

Rotational energies are thus in the far infrared region. Finally we have translational energy that provides a reservoir into which energy from the other molecular motions can flow. The translational energy per molecule is not governed by intrinsic molecular parameters, but is given by

$$E_{\text{trans}} \simeq \tfrac{3}{2} kT \qquad (3.38)$$

For low-frequency vibrations or at high temperatures, translational energies can be comparable to vibrational energies. As we say above, the amount of energy transferred plays an important role in the energy transfer probability. In general, it is the ratio $\Delta E_{\text{trans}} / E_{\text{trans}}$ that is important, where ΔE_{trans} is the energy transferred into or out of translational motion, rather than the total energy transferred. Thus the rate for a purely vibrational→vibrational energy transfer may be significantly different than that for a transfer that involves translational energy as well.

Molecular energy transfer processes are nonadiabatic (Herzfeld and Litovitz, 1959; Landau and Teller, 1936) in that the molecular quantum numbers undergo a change during the energy transfer. This then means that the energy transfer during a collision must occur in a time shorter than a vibrational period.

If we define the interaction or collision time by τ_c, the period of vibration by t_v, and $l = u\tau_c$ as the characteristic length of interaction for a relative velocity u between the two molecules, then the probability of energy transfer per gas kinetic collision varies as (Landau and Teller, 1936)

$$p_{ij} \sim e^{-\omega l / u} \qquad (3.39)$$

where $\omega = 2\pi / t_v$ and is the vibrational angular frequency. Assuming the gas is at thermal equilibrium at temperature T, the average molecular velocity is

$$u = \left(\frac{\omega l k T}{\mu}\right)^{1/3} \qquad (3.40)$$

TABLE 3.1   Comparative Order of Magnitudes for p_{ij} for Various Energy-Transfer Processes

Process		p_{ij} (relative)
Rot–rot	(resonant)	10–0.2
Rot–rot	(nonresonant)	0.2–0.05
Vib–vib	(resonant)	0.2–0.05
Vib–rot	(resonant)	0.1–0.01
Rot–trans		0.1–0.01
Vib–vib	(nonresonant)	0.05–0.001
Vib–rot	(nonresonant)	0.01–0.001
Vib–trans		0.01–0.0001

where μ is the reduced mass of the colliding system. Thus for relaxation processes involving strong, short-range potentials, the energy transfer rate varies as

$$p_{ij} \sim \exp - \left[\frac{\mu (l\,\Delta E)^2}{\hbar^2 kT} \right]^{1/3} \tag{3.41}$$

where ΔE is the vibrational→translational energy transferred.

Although (3.41) was formulated for vibrational→translational, energy transfers, it also holds for electronic→translational, vibrational→vibrational, and rotational→rotational processes as well. We note that p_{ij} increases substantially as ΔE decreases, and in particular there are resonance effects if ΔE becomes very small. This can be the case for rotational→rotational and vibrational→vibrational processes in which small amounts of the total energy transferred actually go into translational energy. In addition to resonance effects, long-range forces must also be included, such as those arising from electric quadruple interactions.

Table 3.1 presents the order of magnitudes of the various types of energy transfers. We note that for rotational→rotational relaxation, the probability may be greater than one. This is possible because long-range forces may make the inelastic scattering cross section larger than the elastic cross section (Cross and Gordon, 1966).

3.5   CONCLUSIONS

From a knowledge of the energy transfer probability p_{ij}, we can determine the collisional deexcitation rate c_{ij} and then evaluate the collisional deexcitation lifetime τ_c. At this point, we are then able to compute the photo-acoustic signal q using (3.26). Experimental results agree well with the theory, and it is assumed that the theory of gas photoacoustics is now well understood. For more information about signal generation in gas photoacoustic systems, we refer the reader to a book edited by Pao (Pao, 1977).

REFERENCES

Cross, R. J., Jr., and Gordon, R. G. (1966). *J. Chem. Phys.* **45**, 3571.

Herzfeld, K. F., and Litovitz, T. A. (1959). *Absorption and Dispersion of Ultrasonic Waves*, pp. 260–348, Academic Press, New York.

Kreuzer, L. B. (1977). In *Optoacoustic Spectroscopy and Detection*, (Y.-H. Pao, Ed.), p. 2, Academic Press, New York.

Landau, L., and Teller, E. (1936). *Phys. Z. Sowjetunion* **10**, 34.

Pao, Y.-H. (Ed.) (1977). *Optoacoustic Spectroscopy and Detection*, Academic Press, New York.

Pao, Y.-H., and Claspy, P. C. (1975). "An Investigation of the Feasibility of Use of Laser Optoacoustic Detection for the Detection of Explosives," Case Western Reserve University, Final Report for Subcontrast 44343-V, The Aerospace Corp.

Stettler, J. D. and Witriol, N. M. (1977). In *Optoacoustic Spectroscopy and Detection*, (Y.-H. Pao, Ed.), p. 29, Academic Press, New York.

Taylor, R. L., and Bitterman, S. (1969). *Rev. Mod. Phys.* **41**, 26.

4

GAS PAS SYSTEMS

4.1 INTRODUCTION

Gas photoacoustic systems have been in continual development and refinement since Viengerov's work of 1938. At present there are a great many different designs, with perhaps their only common feature being the use of a microphone to detect the signal. In this chapter we review some of the basic concepts utilized in the design of a photoacoustic cell for gaseous studies and describe a few nonresonant and resonant cells. We also discuss the design criteria that need to be considered in constructing a complete PAS system.

4.2 THE ACOUSTIC SIGNAL

An acoustic disturbance in a gas can be described by an acoustic pressure $p(r, t)$, which is the difference between the total pressure P and its average value P_0.

$$p = P - P_0 \qquad (4.1)$$

Associated with the acoustic pressure p is an acoustic velocity $u(r, t)$ and an acoustic temperature $\theta(r, t)$. The acoustic velocity is the sound-induced fluid velocity, while the acoustic temperature is the sound-induced change in gas temperature.

The heat $H(\vec{r}, t)$ produced by the absorption of light generates the acoustic signal (Morse and Ingard, 1961). This can be described by

$$\nabla^2 p - \frac{1}{c_0{}^2} \frac{\partial^2 p}{\partial t^2} = -\left[\frac{(\gamma - 1)}{c_0{}^2} \right] \frac{\partial H}{\partial t} \qquad (4.2)$$

where c_0 is the velocity of sound and $\gamma = C_p / C_v$, the ratio of the specific heats of the gas. The above equation does not explicitly include the loss terms due to viscosity and thermal conduction. These loss terms are included later as a perturbation.

Equation (4.2) is an inhomogeneous wave equation that may be solved by taking the time Fourier transform on both sides and expressing the solution p as an infinite series expansion of the normal mode solutions p_j of the homogeneous wave equation. Taking the Fourier transform of (4.2) we have

$$\left(\nabla^2 + \frac{\omega^2}{c_0{}^2}\right)p(\vec{r}, \omega) = \left[\frac{(\gamma - 1)}{c_0{}^2}\right]i\omega H(\vec{r}, \omega) \qquad (4.3)$$

and

$$p(\vec{r}, t) = \int p(\vec{r}, \omega)e^{-i\omega t}\,d\omega \qquad (4.4)$$

$$H(\vec{r}, t) = \int H(\vec{r}, \omega)e^{-i\omega t}\,d\omega \qquad (4.5)$$

The normal mode solutions of the homogeneous wave equation are determined by the boundary conditions. Since the walls of the photoacoustic cell are rigid, the component of the acoustic velocity normal to the walls equals zero at the walls. The acoustic velocity is related to the pressure through the expression

$$\vec{u}(\vec{r}, \omega) = \frac{1}{i\omega\rho_0}\overline{\nabla}\cdot p(\vec{r}, \omega) \qquad (4.6)$$

where ρ_0 is the average gas density. Thus $\overline{\nabla}\cdot p$ must equal zero at the walls. This boundary condition determines the normal mode solutions p_j of the homogeneous wave equation

$$\left(\nabla^2 + k_j{}^2\right)p_j(r) = 0 \qquad (4.7)$$

with

$$k = \frac{\omega}{c_0} \qquad (4.8)$$

Let the normal modes $p_j(r)$ have the resonant frequency ω_j. Since the normal modes are orthogonal, they obey the relationship

$$\frac{1}{V_c}\int p_i^* p_j\,dV = \delta_{ij} \qquad (4.9)$$

where V_c is the volume occupied by the gas.

When the photoacoustic cell is a simple cylinder of radius a and length l (4.7) can be rewritten in cylindrical coordinates (r, θ, z)

$$\frac{1}{r}\frac{\partial}{\partial r}\left(r\frac{\partial p_j}{\partial r}\right) + \frac{1}{r^2}\frac{\partial^2 p_j}{\partial \theta^2} + \frac{\partial^2 p_j}{\partial z^2} + k_j{}^2 p_j = 0 \qquad (4.10)$$

The solution (Morse, 1948) of this equation is given by

$$p_j = \frac{\cos}{\sin}(m\phi)\left[AJ_m(k_r r) + BN_m(k_r r)\right]\left[C\sin(k_z z) + D\cos(k_z z)\right]$$

$$(4.11)$$

where J_m and N_m are Bessel functions of the first and second kind, respectively. For a cylinder, $B=0$ since $N_m(0) = \infty$. To satisfy our boundary conditions, the gradient of p normal to the walls must equal zero. Thus

$$\left(\frac{\partial p_j}{\partial z}\right)_{z=0,l} = 0 \quad \text{and} \quad \left(\frac{\partial p_j}{\partial r}\right)_{r=a} = 0 \qquad (4.12)$$

This then sets $C=0$, and

$$k_z = \left(\frac{\pi}{l}\right)n_z, \qquad n_z = 1,2,3\ldots \qquad (4.13a)$$

$$k_r = \frac{\pi\alpha_{mn}}{a} \qquad (4.13b)$$

where α_{mn} is the nth root of the equation involving the mth order Bessel function. Thus

$$p_j = \frac{\cos}{\sin}(m\phi)\left[AJ_m(k_r r)\right]\left[D\cos(k_z z)\right] \qquad (4.14)$$

Putting this solution for p_j into (4.10) and setting $k_j = \omega_j/c_0$, we obtain for the resonant frequency ω_j,

$$\omega_j = c_0\left(k_z{}^2 + k_r{}^2\right)^{1/2} \qquad (4.15)$$

The normal modes can be divided into pure longitudinal and pure radial. For example, the lowest order longitudinal mode has $n_z=1$, $m=0$, $n=0$, while the lowest order radial has $n_z=1$, $m=1$, $n=0$. As an indication of the order of magnitude of these resonance frequencies, let us consider a typical photoacoustic cylinder with $l = 7$ cm, $a = 0.5$ cm. Taking $c_0 = 3.3 \times 10^4$ cm/sec we obtain 2.37 kHz for the lowest order longitudinal mode and 19.4 kHz for the lowest order radial mode.

The acoustic pressure p within the cell is simply the sum over all the normal modes.

$$p(\vec{r}, \omega) = \sum_j A_j(\omega) p_j(\vec{r}) \qquad (4.16)$$

Substituting (4.16) into (4.2) and making use of (4.7)–(4.10) gives for the mode amplitudes A_j,

$$A_j(\omega) = -\frac{i\omega}{\omega_j^2} \frac{[(\gamma - 1)/V_c] \int p_j^* \, H dV}{1 - \omega^2/\omega_j^2} \qquad (4.17)$$

The numerator of (4.17) represents the coupling between the heat source and the normal mode p_j. The denominator gives the resonance condition for the mode p_j.

So far we have not included any losses from fluid viscosity and heat conduction. We may do so by modifying (4.17) to include a mode damping described by the quality factor Q_j. Thus

$$A_j = -\frac{i\omega}{\omega_j^2} \frac{[(\gamma - 1)/V_c] \int p_j^* \, H dV}{1 - (\omega/\omega_j)^2 - i(\omega/\omega_j Q_j)} \qquad (4.18)$$

If we assume that the optical beam suffers only a small attenuation traversing the cylinder, then $H = \beta I(\vec{r}, \omega)$, where β is the absorption coefficient (cm^{-1}) and $I(\vec{r}, \omega)$ represents the optical beam intensity (ergs/cm²-sec). Under these conditions (4.18) becomes

$$A_j(\omega) = -\frac{i\omega}{\omega_j^2} \frac{\beta[(\gamma - 1)/V_c] \int p_j^* I \, dV}{1 - (\omega/\omega_j)^2 - i(\omega/\omega_j Q_j)} \qquad (4.19)$$

Let us consider two special cases. If I is spatially constant, that is,

$$I(\vec{r}, \omega) = I(\omega)$$

then

$$\int p_j * I \, dV = 0 \qquad \text{for } j \neq 0$$

The only nonzero mode is p_0 with a resonant frequency $\omega_0 = 0$. This is thus a spatially independent pressure change in the cell, independent of r. We then find that

$$A_0(\omega) = \frac{i\beta(\gamma - 1)I}{\omega[1 - i/\omega\tau_0]} \qquad (4.20)$$

where τ_0 is the damping time of p_0 resulting from heat conduction from the gas to the walls of the cell.

Similarly, it can be shown that if the first-order mode p_1 is excited, then

$$A_1(\omega) = -\frac{i\omega}{\omega_1^{\,2}} \frac{\beta(\gamma - 1)I}{1 - (\omega/\omega_1)^2 - i(\omega/\omega_1 Q_1)} \qquad (4.21)$$

If we set the beam power equal to W (W/cm^2), then $I = Wl/V_c$, where l is the length of the cylinder and V_c is the cell volume. Then (4.20) and (4.21) become

$$A_0(\omega) = \frac{i\beta(\gamma - 1)Wl}{\omega(1 - i/\omega\tau_0)V_c} \qquad (4.22)$$

$$A_1(\omega) = -\frac{i\omega}{\omega_1^{\,2}} \frac{\beta(\gamma - 1)Wl}{\left[1 - (\omega/\omega_1)^2 - i(\omega/\omega_1 Q_1)\right]V_c} \qquad (4.23)$$

Comparing the zero resonance and first resonance signals, we have

$$\frac{A_1(\omega_1)}{A_0(0)} = \frac{Q_1}{\omega_1 \tau_0} \qquad (4.24)$$

At low frequencies, it is clear that $A_0(\omega) > A_1(\omega)$; at higher frequencies $A_0(\omega) \propto (1/\omega)$ while $A_1(\omega)$ reaches a maximum at $\omega = \omega_1$. Also $A_1(\omega_1) \propto 1/\omega_1$ and since the first radial resonance frequency $\omega_1 \propto (1/a)$ it is apparent that small-diameter cylinders have a high ω_1, and thus a low $A_1(\omega_1)$, and should therefore be used in the nonresonant $A_0(\omega)$ mode. Also at low frequencies $A_0(\omega) \simeq \tau_0 \beta(\gamma - 1)(Wl/V_c)$ and thus is relatively independent of ω. Thus at low frequencies, the nonresonant $A_0(\omega)$ mode would be the preferred operating condition.

4.3 CELL DESIGN FOR GAS PAS SYSTEMS

Figure 4.1 is a schematic drawing showing a number of optical configurations that have been utilized in gas photoacoustic spectrometers. The single-pass nonresonant cell is a simple cylindrical cell with windows at both ends. The optical beam passes through the cell once along the cylinder axis, and the cell is operated in the nonresonant $A_0(\omega)$ mode. The single-pass resonant cell is usually considerably larger in radius and operates in the $A_1(\omega_1)$ mode, where ω_1 is the first radial resonance. Radial resonances are preferred over longitudinal resonances, since ω_1 for a radial resonance is usually considerably smaller than the equivalent ω_1 for a longitudinal resonance, assuming the same dimension in both cases. The main disadvantage of a resonance cell is that its volume V_c is generally considerably greater than that of a nonresonant cell. Thus it generally requires a fairly high Q_1 to overcome the problems of the greater volume and higher frequency associated with a resonant cell.

In the resonant multipass system the windows are internally reflective so that once an optical beam enters the cell it will bounce back and forth between the two reflective end windows, resulting in several passes through the gas. This effectively increases the interaction length between the optical beam and the gas molecules without increasing the actual cell volume (Kamm, 1973; Goldan and Goto, 1974).

Another method to increase sensitivity is to place the photoacoustic cell within the laser cavity, a logical extension of the use of intracavity absorption cells in gas lasers (Chakerian and Weisbach, 1973). As

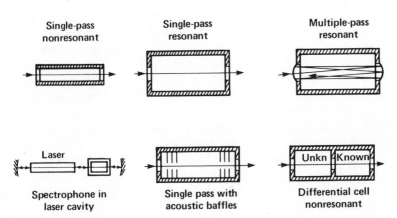

Figure 4.1 Schematic drawing of various designs for photoacoustic gas cells. (Reproduced by permission from Dewey, C. F., Jr., 1977.)

Viengerov discovered in 1938, one of the major limitations to the sensitivity of gas photoacoustic systems is the background photoacoustic signals that arise from optical heating of the cell windows. One method around this problem is to use a single-pass nonresonant cell containing acoustic baffles near the windows. This configuration is, of course, inappropriate for resonant systems since the Q would be greatly reduced (Dewey, 1974; Bruce et al., 1976).

Another method, used by Deaton et al. (1975), involves balancing out the background signal in a differential nonresonant cell. This technique is similar to that used by Luft (1943). Two identical cells are illuminated by the same optical beam, which is assumed to undergo only a small amount of absorption in passing through the windows and gases of the two cells. If both cells contain a nonabsorbing gas, then only background signals are produced and the differential signal is zero. With an absorbing gas in the sample cell, a differential signal essentially devoid of background is obtained.

Finally there is a cell design that combines several of the attractive features of some of the other designs. This is a multiple-pass resonant cell (Kamm and Dewey, 1973; Goldan and Gotto, 1974). In this design the reflective surfaces of the windows are located on the exterior and thus do not contribute to the acoustic background of the system.

4.4   NOISE IN GAS PAS SYSTEMS

4.4.1   Acoustic Noise

The most serious limitation to high sensitivity of gas PAS systems has been the background signal that is due to the photoacoustic signal from absorption of the optical beam in the cell windows, and to a lesser degree from absorption of scattered radiation by the cell walls. The effects of window heating can be diminished by using a differential cell design, as is described earlier. Also, increasing the cell chamber length should result in a lower window background signal as compared to the gas signal.

Another successful method has been to subtract the window signal from the total signal. This is relatively straightforward since the window absorption is almost independent of wavelength in the spectral regions usually used for gas photoacoustic studies. Alternatively, it is possible to modulate the wavelength about the center of the gas absorption line. This technique has been used very successfully by Patel (1973) and Patel et al. (1974).

Additional background acoustic noise can arise from ambient acoustic and building vibrations and from electromechanical light-chopping systems. Good cell design and the use of a phase-sensitive lock-in amplifier significantly lessen the noise due to ambient acoustic and mechanical vibrations. Chopper noise can be diminished by proper acoustic and vibrational insulation between the chopper and the photoacoustic cell. In some cases, PAS studies are performed on flowing gas samples. If the flow of the gas into the chamber is turbulent, very strong background signals are produced. This background noise can be minimized by decreasing the turbulence of the flow and by introducing the sample gas at a node of the standing wave of a resonant mode.

4.4.2 Electronic Noise

The electronic noise in a gas photoacoustic system is mainly due to noise sources in the amplifier connected to the microphone (Kreuzer, 1971). There are three noise sources in the amplifier; a series voltage noise source, a shunt current noise source, and resistor Johnson noise. Amplifier voltage noise is essentially frequency independent. Amplifier current noise and resistor Johnson noise decrease with increasing frequency. Thus at low frequencies, Johnson noise and amplifier current noise dominate, while voltage noise is the predominate term at high frequencies.

4.4.3 Brownian Motion Noise

The Brownian motion, or thermal fluctuations, of the gas in the photoacoustic cell is itself the ultimate limit to the sensitivity of a gas PAS spectrometer. These thermal fluctuations can excite acoustic normal modes in the cell, and thus fundamentally limit photoacoustic sensitivity. Kittel (1958) has shown that the jth normal mode excited by the Brownian motion has an amplitude given by

$$|A_j(\omega)|^2 = \frac{4\rho_0 c_0{}^2 kT}{V_c \omega_j Q_j \left[\left(1 - \omega^2/\omega_j{}^2 \right)^2 + \left(\omega/\omega_j Q_j \right)^2 \right]} \qquad (4.25)$$

For $\omega \ll \omega_j$ the resonant frequency of mode j,

$$|A_j(\omega)|^2 \simeq \frac{4\rho_0 c_0{}^2 kT}{\omega_j Q_j V_c} \qquad (4.26)$$

4.4.4 Microphone Noise

A condenser microphone employs a metal or metalized foil or diaphragm mounted under a large radial tension. Acoustic pressure acting on one side of the diaphragm causes motion, thereby changing the electrical capacitance between the diaphragm and a fixed metal plate located a small distance behind the diaphragm. The motion of the diaphragm can be described in terms of the normal modes of vibration of a thin plate supported around its perimeter. Only the lowest-order mode of vibration actually causes a significant change in capacitance, and that mode is the one that causes the diaphragm to bend toward a spherical shape. In this mode of vibration there is a displacement of each point of the diaphragm in the x direction by an amount given by

$$x(r) = x(0)\left(1 - \frac{r^2}{r_0^{\,2}}\right) \tag{4.27}$$

where r is the distance of the point of interest from the center of the diaphragm, $x(0)$ is the displacement of the center point, and r_0 is the radius of the diaphragm. The average displacement of the diaphragm is then given by

$$\bar{x} = \frac{1}{\pi r_0^{\,2}} \iint x(r) r \, dr \, d\theta = \tfrac{1}{2} x(0) \tag{4.28}$$

The equation of motion in the average coordinate $\bar{x}$ for the lowest order mode is

$$m \frac{d^2 \bar{x}}{dt^2} + \zeta \frac{d\bar{x}}{dt} + \chi \bar{x} = F \tag{4.29}$$

where m is the mass of the diaphragm, ζ is the damping factor, χ is the restoring force, and F is the driving force. The driving force is due to the acoustic force, $p \pi r_0^{\,2}$, where p is the acoustic pressure, and to the force due to the presence of an external voltage bias.

We can define a microphone equivalent volume V_m such that

$$V_m = \frac{\gamma P_0 (\pi r_0^{\,2})^2}{8 \pi T_m} \tag{4.30}$$

where T_m is the tension in the diaphragm. We can regard V_m as a volume enclosed by the diaphragm that acts as the restoring force χ when compressed ($\chi = 8 \pi T_m$).

When the microphone is connected to the input of a high impedance amplifier, then the output voltage V_s is given by (Olson, 1947)

$$V_s = p \frac{V_B \pi r_0^2}{d\chi} \left(1 - \frac{\omega^2}{\omega_m^2} - i\frac{\omega}{\omega_m Q_m} \right)^{-1}$$ (4.31)

where ω_m is the microphone resonance frequency, $\omega_m = (\chi/m)^{1/2}$, and Q_m is the microphone quality factor, $Q_m = (m\chi/\zeta)^{1/2}$, V_B is the microphone bias voltage, and d is the capacitive spacing in the microphone. The open-current voltage sensitivity is defined as the low-frequency ($\omega \ll \omega_m$) ratio V_s/p. Thus

$$S = \frac{V_s}{p} = \frac{V_B \pi r_0^2}{d\chi}$$ (4.32)

$$S = \frac{V_B V_m}{d\gamma P_0 \pi r_0^2}$$ (4.33)

If the photoacoustic cavity is small enough and the operating frequency is much less then either the cell's first resonant frequency or the microphone's resonant frequency, then the microphone can be considered as being simply an added volume V_m to the cell volume V_c. Then the volume V_c in (4.22) is replaced by $(V_c + V_m)$ and the zeroth-order mode has an amplitude

$$A_0(\omega) = \frac{i\beta(\gamma - 1)Wl}{\omega(1 + i/\omega\tau_0)(V_c + V_m)}$$ (4.34)

For $\omega\tau_0 \gg 1$ the voltage signal from the microphone due to the absorption of light is given by

$$V_s = pS$$

$$= \frac{i(\gamma - 1)\beta Wl}{\omega(V_c + V_m)} \cdot S$$ (4.35)

or

$$V_s = \frac{i(\gamma - 1)V_B}{\omega\gamma P_0 \pi r_0^2 d} \left(\frac{V_m}{V_c + V_m} \right) \beta Wl$$ (4.36)

In addition to the photon-generated signal, there is the noise due to Brownian motion. When the photoacoustic cell is small enough, the effective restoring force of the microphone is altered, since the movement of the diaphragm is also hindered by compression of the gas in the cell itself. Thus

$$\chi' = \chi \left[1 + \frac{V_m}{V_c} \right] \tag{4.37}$$

The microphone resonant frequency is also altered

$$\omega'_m = \omega_m \left[1 + \frac{V_m}{V_c} \right]^{1/2} \tag{4.38}$$

and the Brownian noise signal is given by

$$|V_{sn}(\omega)|^2 = p_n{}^2 S^2$$

$$= |A_n(\omega)|^2 S^2$$

$$= \frac{4\rho_0 c_0{}^2 kT}{\omega_m Q_m V_m (1 + V_m/V_c)^2} \cdot S^2 \tag{4.39}$$

If we now consider the signal/noise ratio

$$\left(\frac{s}{n} \right)^2 = \frac{|V_s(\omega)|^2}{|V_{sn}(\omega)|^2}$$

$$= \frac{(\gamma - 1)^2}{4kT\omega^2 \gamma P_0} \left(\frac{l}{V_c} \right)^2 V_m \omega_m Q_m \beta^2 W^2 \tag{4.40}$$

Since

$$V_m = \frac{\gamma P_0 (\pi r_0{}^2)^2}{\chi}$$

Thus

$$\left(\frac{s}{n} \right)^2 = \frac{(\gamma - 1)^2}{4kT\omega^2} \left(\frac{l\pi r_0{}^2}{V_c} \right)^2 \left(\frac{\omega_m Q_m}{\chi} \right) \beta^2 W^2 \tag{4.41}$$

The factor $(\gamma - 1)^2$ can be increased to some extent by going to a monatomic gas such as helium. Also (s/n) can be improved by operating at lower temperatures. The factor $(l\pi r_0{}^2/V_c)^2$ may be regarded as a coupling coefficient between the microphone and the photoacoustic cavity. Clearly, this coupling increases for larger microphones (larger r_0) and small cross-sectional areas of the cell. In particular, for nonresonant cells, we would prefer to keep the cross-sectional area of the cell only slightly larger than the area of the optical beam.

Furthermore (4.41) indicates that (s/n) increases as ω decreases, provided, of course, that $\omega > 1/\tau_0$. It should also be kept in mind that at lower frequencies the $1/f$ Johnson noise of the microphone amplifier may become significant.

Calculations performed by Kreuzer (1977) indicate that a good cell design employing a sensitive microphone and associated electronics gives a noise equivalent power (NEP) when $(s/n)^2 = 1$, in air at room temperature, of

$$\text{NEP} \sim 10^{-11} \ W/\sqrt{\text{Hz}}$$

for a cell 30 cm long and 1 cm in diameter. The absorbed power Wl gives a minimum detectable absorption $\beta_{\min}$ of

$$\beta_{\min} \sim 10^{-12} \ \text{cm}^{-1} W/\sqrt{\text{Hz}}$$

4.5 RESONANCE CONDITIONS

In (4.17), there is no resonance quality factor associated with the resonant frequency ω_j. This results in a physically unreasonable situation where $A_j(\omega)$ goes to infinity as $\omega \to \omega_j$. Realistically, a finite Q_j is present as in (4.18). This finite Q is basically the result of acoustic losses due to heat conduction and viscosity.

Heat conduction and viscosity losses can be separated into a surface loss and a volume loss (Morse and Ingard, 1961). The surface loss occurs in a thin region near the walls. This region may be considered to be composed of two layers, one of thickness l_v, in which the viscosity effects takes place, and one of thickness l_h, in which heat conduction takes place. These skin depths are given by

$$l_v = \left(\frac{2\eta}{\omega \rho_0} \right)^{1/2} \tag{4.42}$$

$$l_h = \left(\frac{2\kappa}{\rho_0 \omega C_p} \right)^{1/2} \tag{4.43}$$

where η is the viscosity, κ is the thermal conductivity, ρ_0 is the density, and C_p is the specific heat at constant pressure of the gas in the photoacoustic cell. The acoustic frequency is given by ω.

The physical nature of these two losses can be understood by the boundary effects imposed by the walls. Since the thermal conductivity of the walls is greater than that of the gas, the gas near the walls is kept at constant temperature, and thus the expansion and contraction associated with the acoustic wave occur under isothermal conditions. On the other hand, the expansion and contraction of the gas far from the walls is nearly adiabatic. Acoustic loss from heat conduction occurs in the region where the gas behavior is partly adiabatic and partly isothermal.

Similarly, at the wall surfaces, the tangential component of acoustic velocity is zero because of the viscosity, while far from the wall it is proportional to the gradient of the acoustic pressure. Viscoelastic loss occurs in the region where the tangential velocity approaches zero. The total surface loss L_s is given by

$$L_{sj} = |A_j|^2 \int \left(\tfrac{1}{2} R_v |v_{tj}|^2 + \tfrac{1}{2} R_h |p_j|^2 \right) dS \qquad (4.44)$$

where

$$R_v = \left(\tfrac{1}{2} \eta \omega \rho_0 \right)^{1/2}, \qquad (4.45)$$

$$R_h = \frac{\gamma - 1}{\rho_0 c_0{}^2} \left(\frac{\kappa \omega}{2 \rho_0 C_v} \right)^{1/2} \qquad (4.46)$$

and v_t is the tangential velocity far from the walls.

The volume loss is due simply to energy transferred from the acoustic wave to thermal energy through heat conduction and through the viscosity, which acts as a frictional force. Thus

$$L_{vj} = |A_j|^2 \left(\frac{\omega_j}{\rho_0 c_0{}^2} \right)^2 \left[(\gamma - 1) \frac{\kappa}{2 C_p} + \frac{2\eta}{3} \right] V_c \qquad (4.47)$$

The acoustic energy stored in mode j is given by

$$E_j = \frac{V_c |A_j|^2}{\rho_0 c_0{}^2} \qquad (4.48)$$

The Q of the mode is simply defined as

$$Q_j = \omega_j \frac{\text{energy stored in mode } j}{\text{rate of energy loss from mode } j}$$

$$= \omega_j \frac{E_j}{L_{sj} + L_{vj}} \qquad (4.49)$$

Viscous and thermal losses are not the only losses that occur in a photo-acoustic cell. Other losses occur through wave reflection losses resulting from the compliance of the chamber walls. Losses also occur from the compliance of the microphone diaphragm and, in fact, increase as the sensitivity of the microphone increases. In addition, serious losses can occur from acoustic wave scattering at surface obstructions and deviations in chamber geometry from the normalized resonant mode shape.

Since high-sensitivity microphones generally have large surface areas, 2–3 cm in diameter, most gas photoacoustic cells have cylindrical diameters of 2–5 cm. In this region, the first-order radial resonant mode tends to have a lower resonance frequency than does the first-order axial mode for cells less than 10 cm in length. Thus many gas PAS systems are operated in the first-order radial resonance mode. Goldan and Goto (1974) and Max and Rosengren (1974) have achieved values of Q, at this resonance, as high as 770–890 for cells 5–10 cm in diameter. Such high values of Q overcome the decrease in signal from working at frequencies of 2000–6000 Hz, instead of the usual nonresonant frequencies of 50–200 Hz.

In addition to the radial and axial resonant cell designs, there has been considerable interest in the use of cells that employ a Helmholtz resonance. An example of such a design is depicted in Figure 4.2. Here we have two volumes V_1 and V_2 connected by a channel of length l_c.

The Helmholtz resonance can be understood by considering the gas in the channel to form a plunger that can move along the channel. As the plunger moves toward one volume, say V_1, it compresses the gas in V_1 and rarefies the gas in V_2, thereby experiencing a restoring force proportional to its displacement along the channel. There is, in addition, a retarding force on the plunger due to the thermoviscous effects described above. The Helmholtz resonance frequency is given by

$$\omega_H = c_0 \left(\frac{A}{l_c V_r} \right)^{1/2} \qquad (4.50)$$

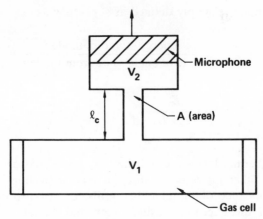

Figure 4.2 Schematic drawing of a photoacoustic gas cell with a Helmholtz resonance configuration.

where

$$V_r = V_1 V_2 / (V_1 + V_2) \qquad (4.51)$$

A is the cross-sectional area, and l_c is the length of the channel. The quality factor Q_H is given by

$$Q_H = \frac{\rho_0 A}{8 \pi \eta} \left[\omega_H{}^2 - \left(\frac{4 \pi \eta}{\rho_0 A} \right)^2 \right] \qquad (4.52)$$

A major advantage of the Helmholtz design is that the resonant frequency can be altered considerably by simply changing the dimensions A and l_c of the channel only. Thus V_1 can be the section of the photoacoustic cell through which the radiation passes, V_2 can be the volume containing the microphone, and l_c can be the channel connecting the two. By a suitable design of the channel dimensions, a resonance condition having a reasonably high Q ($Q \sim 100$) can be achieved with a fairly small total gas volume.

4.6 CONCLUSIONS

As is discussed in this chapter, a gas photoacoustic system can have a considerably enhanced signal when careful cell design is used. In particular, much is to be gained from the use of a resonant design system in many situations. The resonances that can be used are longitudinal, radial, and even Helmholtz resonance modes. Resonance conditions are, of course, not suitable in all situations, and in particular, where total cell volume must be minimized, a nonresonant design is preferable.

REFERENCES

Bruce, C. W., Sojka, B. Z., Hurd, B. G., Watkins, W. R., White, K. O., and Derzko, Z. (1976). *Appl. Opt.* **15**, 2970.

Chakerian, C., Jr., and Weisbach, M. F. (1973). *J. Opt. Soc. Am.* **63**, 342.

Deaton, T. F., Depatie, D. A., and Walker, T. W. (1975). *Appl. Phys. Lett.* **26**, 300.

Dewey, C. F., Jr. (1974). *Opt. Eng.* **13**, 483.

Dewey, C. F., Jr. (1977). In *Optoacoustic Spectroscopy and Detection* (Y.-H. Pao, Ed.), pp. 47–77, Academic Press, New York.

Goldon, P. D., and Goto, K. (1974). *J. Appl. Phys.* **45**, 4350.

Kamm, R. D. (1973). "An Acoustic Amplifier for the Detection of Atmospheric Pollutants," M. S. Thesis, MIT (unpublished).

Kamm, R. D., and Dewey, C. F., Jr. (1973). *Proc. IEEE/OSA Conf. Laser Eng. Appl., May 30–June 1*, 35.

Kittel, C. (1958). *Elementary Statistical Physics*, Part 2, Wiley, New York.

Kreuzer, L. B. (1971). *J. Appl. Phys.* **42**, 2934.

Kreuzer, L. B. (1977). In *Optoacoustic Spectroscopy and Detection* (Y.-H. Pao, Ed.), pp. 1–25, Academic Press, New York.

Luft, K. F. (1943). *Z. Tech. Phys.* **24**, 97.

Max, E., and Rosengren, L.-G. (1974). *Opt. Commun.* **11**, 422.

Morse, P. M. (1948). *Vibration and Sound*, McGraw-Hill, New York.

Morse, P. M., and Ingard, K. V. (1961). In *Encyclopedia of Physics* Vol. XI/1, (S. Flugge, Ed.), Springer-Verlag, Berlin and New York.

Olson, H. F. (1947). *Elements of Acoustical Engineering*, pp. 220–224, Van Nostrand-Reinhold, Princeton.

Patel, C. K. N. (1973). In *Coherence and Quantum Optics* (L. Mandel and E. Wolf, Eds.), pp. 567–593, Plenum Press, New York.

Patel, C. K. N., Burkhardt, E. G., and Lambert, C. A. (1974). *Science* **184**, 1173.

Viengerov, M. L. (1938). *Dokl. Akad. Nauk SSSR* **19**, 687.

5

RADIATION SOURCES

5.1 INTRODUCTION

In photoacoustics it is the absorption of electromagnetic radiation that gives rise to the acoustic signal. It should, however, be kept in mind that a photoacoustic signal could be generated through the absorption of other energies, such as particle absorption through electron or ion beam bombardment. However, we ignore at this time any other sources of energy except electromagnetic and, in fact, restrict our discussions to sources of optical radiation (UV to IR) only.

The radiation sources described in this chapter are those of interest primarily to investigators performing photoacoustic studies on gases. However, most, if not all, of these sources are equally acceptable for use in studies of liquids and solids. When appropriate, the radiation source includes not only the actual photon source, but the monochromator or wavelength dispersion system and the modulation mechanism as well.

As we see later, the parameters of importance for electromagnetic sources for PAS systems are (a) the power available per usable bandwidth; (b) the wavelength range; (c) the tunability of the source; and (d) the ease of modulation of intensity and sometimes of wavelength as well.

5.2 INCOHERENT SOURCES: UV–VISIBLE

In this section we are concerned with conventional incoherent sources of optical radiation in the 0.1–1 μm spectral region. Most of these sources have been reviewed by Cann (1969), Koller (1965), Samson (1967), and McNesby et al (1971). Incoherent sources tend to fall into two major categories: incandescent emitters and arc sources.

The emission from an incandescent source can be approximated by radiation emanating from a blackbody at a given temperature. According to the Stefan-Boltzmann law (Kasper, 1972), the total energy emitted per unit area of a blackbody source is proportional to the fourth power of its temperature. The spectral distribution of blackbody radiation is given by Planck's well-known radiation formula and is shown in Figure 5.1.

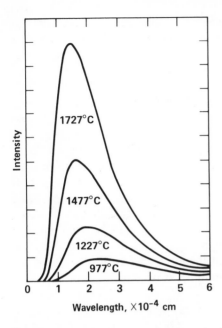

Figure 5.1 Spectral distribution of blackbody radiation at various temperatures.

The initial photoacoustic experiments, performed by Alexander Graham Bell in 1880 and 1881, utilized the least expensive and best-known incandescent source—the sun. The sun provides continuous spectral radiation on earth that ranges from 300 nm to somewhat more than 1 μm (Billings, 1972). Its spectral output closely matches that of a blackbody source operating at 5900°K. At sea level the emission peaks near 500 nm, and wavelengths shorter than 300 nm are effectively blocked by absorption in the atmosphere.

Among man-made sources the tungsten lamp provides one of the simplest and most economical sources of continuous radiation throughout the visible and infrared regions. The standard coiled-coil tungsten filament is contained in a quartz envelope and operates at a temperature of 3000°K. The addition of a halogen gas to the quartz bulb reduces the rate of tungsten vaporization and prolongs the life of the lamp to 1000 hours or more without changing the spectral characteristics. The tungsten lamp can be operated at higher temperatures, but lamp lifetime is rapidly reduced as the melting point of tungsten (3644°K) is approached.

Optical radiation can also be obtained from an arc that occurs when an electrical current is passed through a gas or vapor, causing it to discharge. The gases are usually contained in glass or quartz envelopes at pressures that vary from a few millitors to hundreds of atmospheres. Low-pressure arcs are dominated by radiations at discrete wavelengths characteristic of the particular gas in the discharge. A continuum broad-band spectrum is possible with certain gases under high pressure. This spectrum is the result of both pressure-broadening effects and from excitations of many closely spaced higher energy levels. High-pressure operation can result in high luminous efficiencies. The emission from arc sources both at low and high pressures is quite nonuniform across the arc, with maximum intensity near the cathode and along the arc axis.

One of the oldest sources of optical radiation is the carbon arc lamp, which is created when an open electrical discharge occurs between two carbon electrodes in air. Compared to modern enclosed arcs, the carbon arc is inherently less efficient and has a short lifetime of only a few hours. The carbon arc emission corresponds to a blackbody radiation at 4000–6000°K. Superimposed on the continuum is the strong CN band emission near 388 nm. Emission spectra with somewhat different characteristics can be generated by doping the carbon in the electrodes with other elements. For example, carbon electrodes containing strontium produce arcs with enhanced emission near 700 nm.

A high-pressure mercury lamp is an intense source of multiline radiation between 230 and 650 nm. This source is unfortunately not very useful because its emission spectrum is dominated by lines rather than a continuum. On the other hand, it is extremely intense in the near ultraviolet region 230–400 nm. Furthermore, the small diameter 0.1–5 mm of the arc makes it useful for imagery in optical systems.

The most popular incoherent source in photoacoustic spectroscopy is the high-pressure xenon lamp. This lamp operates at pressures of 50–70 atm and is a very efficient emitter of intense radiation from 230 to 2000 nm. The emission spectrum is primarily a continuum, with only a few intense lines between 800 and 1000 nm. The corresponding color temperature is usually between 8000 and 10,000°K. Some of the commercial xenon lamps, such as those manufactured by Hanovia and Varian, have arc sizes that are quite suitable for imaging in the optical systems used in photoacoustic spectroscopy. Figure 5.2 shows a typical emission spectrum from a high-pressure xenon lamp.

Finding an intense source of radiation below 230 nm is quite difficult, and those sources that do exist cover only a small spectral range. The hydrogen and deuterium arc lamps provide a reasonably intense continuum between 165 and 250 nm. Rare gas capillary lamps (Tanaka, 1955) and microwave electrodeless lamps operating at 2.45 GHz have also provided moderate intensity below 200 nm. A glow-discharge xenon lamp is a fairly weak radiation source in the 180–225 nm range, while the glow-discharge krypton lamp emits in the 125–165 nm range.

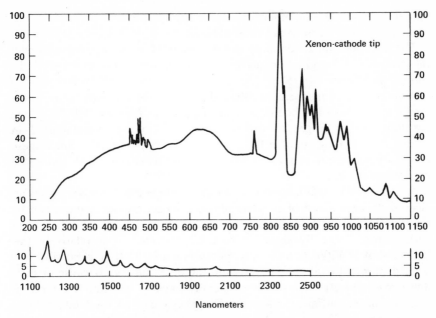

Figure 5.2 A typical emission spectrum from a high-pressure xenon arc lamp.

Thus, in general, there are several good incoherent sources of radiation in the 230–1000 nm range. These sources are usually compact, reliable, rugged, and relatively inexpensive. Emission at wavelengths shorter than 230 nm is less intense and requires several sources to span the 100–230 nm range. In combination with suitable optical systems, monochromatic radiation with bandwidth of 10–100 Å is obtainable with reasonable intensity.

5.3 INCOHERENT SOURCES: INFRARED

Most of the incoherent sources in the infrared region of the optical spectrum are of the incandescent variety. An exception to this is the carbon arc, which can also be used in the UV–visible region.

For the carbon arc, the spectral distribution of the emitted radiation depends on the region of the arc viewed, on the power dissipation, and on the nature of the carbons (Koller, 1965). Typically, the carbon arc spectrum is that of a blackbody operating at 4000–5000°K. The carbon arc can supply fairly intense infrared radiation down to 20 μm. A difficulty with carbon arcs as optical sources for spectroscopy in their short lifetime and the fluctuations in intensity of the arc. The most common use of carbon arcs is as commercial projection arcs.

The high-pressure xenon lamp can also be used in the infrared, although the presence of the glass or quartz envelope limits its use to wavelengths shorter than 3 μm.

A common and inexpensive infrared source is the tungsten lamp. These lamps operate as blackbodies with temperatures of almost 3000°K. They are thus not as bright as carbon arcs, but can be manufactured in configurations more suitable for use as spectroscopic sources. Since these lamps are also contained within glass or quartz envelopes, they cannot be used beyond 3 μm.

A stable source that is used without an envelope is the globar. This is an electrically heated rod of bonded silicon carbide. Globar sources can be made very stable with feedback control of the power supplies to the silicon carbide rod. Typically these sources are operated at a relatively low temperature of 1200°K, although operation above 2200°K is possible if a layer of thorium oxide is sintered onto the outer surface. Because of their low temperature, globar sources generally have to be relatively large in area to ensure sufficient intensity. The globar can be used into the far infrared beyond 30 μm.

A more efficient source for the near-to-mid-infrared regions is the Nernst glower (Hackforth, 1960). This is a cylindrical tube of refractory oxide that is a mixture of zirconia, yttria, and thoria. Since the refractory is nonconducting at room temperature, the tube must first be heated by an auxiliary source before it can be operated. Nernst glowers operate around 1700°K and thus are more intense than globars down to about 14 μm. Beyond 14 μm the globar is more intense, and both operate at about 1200°K beyond 10 μm.

Unlike an ideal blackbody, no real material surface absorbs all the radiation incident on it, and its emissivity is always less than unity. The absorptivity, and hence the emissivity, can be increased by subjecting the incident radiation to multiple reflections. This is the principle of a heated cavity with a small exit aperture. For such cavities, values of emissivity in excess of 0.99 have been obtained. Blackened stainless steel and graphite have been used for the walls of the cavity. Commercial sources may employ conical, recessed conical, or cylindrical cavities. The interior of the cavity may contain an arc source, or an incandescent source. The cavity acts to increase total emissivity and to concentrate the radiation into a smaller exit area.

In general incoherent infrared sources are much less intense than incoherent UV–visible sources. They are also less efficient for spectroscopic studies because their larger surface areas makes imaging into a spectrometer difficult.

5.4   OPTICAL SYSTEMS FOR INCOHERENT SOURCES

For most photoacoustic applications, the incident optical radiation should have a narrow bandwidth, that is, be essentially monochromatic, and be tunable over a wide spectral range. The necessity for monochromatic radiation means that an incoherent source cannot be used by itself but must be used in conjuction with some optical system that filters or disperses the source radiation.

If relatively low resolution can be used (> 10 nm), then the optical system can consist of various types of filters. These optical filters might operate by absorption, reflection, interference, or scattering of the incident light (Calvert and Pitts, 1966). Optical filters are made from glass, gelatin, thin films, plastics, and dissolved organic salts. Discrete tunability can be performed by changing the filter. In addition to the filter, the optical system would most probably also include a lens system suitable for imaging the optical beam into the photoacoustic chamber. For the UV–visible region of the spectrum, these lenses would be fused quartz, which transmits in the 200–2000 nm range, or of optical grade glass that transmits from 350 to 1500 nm.

High-resolution optical spectroscopy in the UV–visible region (< 10 nm) requires the use of a monochromator with the incoherent source. Most monochromators utilize either a prism or a diffraction grating as the dispersive element. The resolution is determined primarily by the dispersing capability of the prism or grating (for a grating this is dependent on the number of groves per millimeter ruled onto the grating blank), and by the width of the entrance and exit slits in the monochromator. Resolutions of 1–5 nm are routinely achievable with most inexpensive monochromators, and resolutions of 1–10 Å can be obtained with the larger high-resolution monochromators.

Tunability with a monochromator is readily achieved by simply rotating the prism or grating. Unlike a filter system, a monochromator can provide a continuous tuning over the full range of the dispersive element. In the case of conventional reflection gratings, more than one grating is often needed to efficiently cover a wide wavelength range. This is a result of the fact that the grooves or lines of the grating are ruled at an angle that maximizes the reflection and dispersion for a particular wavelength range. Along with the grating one also requires a filter or filters to prevent that radiation that is dispersed in higher orders from the grating (Born and Wolf, 1965) from entering the photoacoustic cell.

Since the strength of a photoacoustic signal is directly proportional to the intensity of optical radiation that enters the photoacoustic cell, it is apparent that the optical source should be very bright, that is, have a high spectral radiance, and that the monochromator should be quite fast, that is, have a high light throughput (low f-number). The spectral radiance of a source is defined as the power radiated per unit area per solid angle and per unit optical bandwidth. It is an invariant of the optical system and thus the brightness of the image may never exceed that of the source itself (Born and Wolf, 1965). The spectral radiance of the incoherent sources discussed above range from 10 mW/mm^2-sr-nm for the sun, and for 1 kW xenon, Hg, and C arc lamps, to 1 mW/mm^2 sr-nm for a 1 kW tungsten lamp, 0.1 for a H$_2$ lamp, and 10^{-2} and 10^{-3} for xenon discharge and krypton discharge lamps, respectively.

As in the UV–visible spectral region, monochromatic radiation can be obtained in the infrared region with the use of filters or dispersive optical elements.

The filters may operate on absorption, reflection, light-scattering, or interference principles. The filters may be made of plastics containing dyes, colored glass, or sublimated dyes on glass. Semiconductors such as silicon and germanium are in common use as long-wavelength filters in the infrared, transmitting below their bandgap wavelength and absorbing above it. Interference filters can be used to transmit only a narrow wavelength region. The same is true of Christiansen filters (Wolfe, 1965). These filters are made of small, closely packed particles of an infrared-transparent substance suspended in a liquid or a gas. The optical properties of the materials are so chosen that the indices of refraction of the particles and the suspending medium are the same at the wavelength that is to be transmitted. The $dn/d\lambda$ values of the liquid and the solid particles are chosen to be as widely different as possible. Thus as the wavelength is progressively increased or decreased from the wavelength at which equality of the indices occurs, the difference in the indices creates strong scattering of the light.

For higher resolution work, dispersive elements such as refraction prisms and diffraction gratings are used. These are generally used in monochromators. Since infrared lenses that are usable over a wide spectral range are difficult to obtain, off-axis parabolic mirrors are commonly employed in these monochromators. Infrared monochromators generally have longer focal lengths that UV–visible monochromators because of the longer optical wavelengths.

In the far infrared, the beam exiting from a monochromator is of such low intensity that interference monochromators are used to increase throughput. The larger throughput occurs because the interferometer has a large entrance aperture determined by the mirror size. This enables the instrument to accept more radiant energy from the source than prism or grating instruments, in which the entrance aperture is limited by narrow slits. High sensitivity gain is due to the examination by the instrument of each wavelength throughout the entire time period of each scan. In a conventional dispersion instrument, each wavelength is examined for only $1/n$ of the scan time if n is the number of resolution elements. Thus for the same scan time, the interferometer has a gain of $\sqrt{n}$, which typically can be of the order of 50.

If a photoacoustic cell is placed at the output of an infrared interferometer, then the acoustic signal obtained must be wave analyzed to recover the infrared absorption spectrum (Jacquinot, 1960).

5.5 COHERENT SOURCES: UV–VISIBLE

As we see in the preceding sections, the spectral radiance of conventional UV–visible sources ranges from 10 to as low as 10^{-3} mW/mm²-sr-nm. A coherent radiation source such as a 20-mW dye laser can have a spectral radiance of 10^{10} mW/mm²-sr-nm. Such a high spectral radiance comes from the extremely narrow spectral linewidths and narrowly collimated light beam associated with laser radiation.

In its simplest form, a laser is an optical oscillator, which, like an electronic oscillator, exhibits gain and feedback characteristics. Gain occurs at optical frequencies in the active medium if the laser has two states between which optical transitions are allowed, and if the upper state has a population momentarily greater than that of the lower state (population inversion). The two states involved can be electronic, vibrational or even rotational states. Laser emission has been observed in solids, gases, and liquids, and in the spectral region ranging from the ultraviolet to the far infrared.

To create a population inversion, the active medium of the laser is pumped, that is, the upper level is populated, by one of several means, such as electrical discharge, optical excitation with lamps or other lasers, collisional processes with excited molecules, exothermic chemical reactions, and electron beam excitation.

Once the deexcitation from the upper levels begins, further radiative deexcitation can be stimulated by creating the proper feedback conditions.

This is done by placing the active medium inside an optical resonator or cavity formed by reflective mirrors that entrap the radiation within the cavity. One of the mirrors is partially transmitting so as to couple energy out of the resonator. Positive feedback, of course, does not occur until threshold is reached, that is, until the optical gain exceeds the losses within the cavity. The primary resonator loss mechanisms are absorption and scattering processes by the components of the resonator, and transmission losses through the output mirror.

In general, a laser medium experiences gain over only a small spectral range, usually determined by the width of an absorption line. Within this spectral range there may be several narrower laser oscillations. These oscillations have high gain since they undergo constructive interference within the laser cavity. That is, they satisfy the relationship

$$l = \frac{m\lambda}{2} \qquad (5.1)$$

where l is the laser cavity length, m is an integer, and λ is the laser wavelength. The resonant frequencies thus are given by

$$\nu = \frac{mc}{2l} \qquad (5.2)$$

where c is the velocity of light.

Low-pressure gases and vapors have narrow natural absorption line-widths, and thus usually only one cavity mode exists with sufficient gain to reach threshold. Other materials such as dyes and solids can support many cavity modes because of their relatively broad natural linewidths.

The minimum theoretical bandwidth of a single mode is of the order of 1 Hz or 10^{-9} Å for a visible laser over a 1-sec duration. The observed linewidth is considerably broader because of variations in the optical length of the laser cavity from mechanical vibrations and thermal fluctuations. In practice, laser linewidths of the order of 10^{-4} Å are readily achievable, and with frequency-stabilized lasers, linewidths as narrow at 10^{-9} Å have been attained.

Besides the extremely high spectral purity of a laser emission, the laser beam also displays high spatial coherence, which determines how tightly the beam can be focused and how well it can be collimated. The spatial coherence arises because the laser mode within the cavity is quantized in the transverse directions as well as in the longitudinal direction. The most important transverse mode is the fundamental TEM_{00} mode. This mode has its energy concentrated near the resonator axis and it suffers the smallest diffraction loss upon passage through optical components of finite dimensions. The beam intensity in the transverse direction exhibits a Gaussian profile. The beam radius is the distance at which the field amplitude is e^{-1} times that at the axis, that is, the distance at which the intensity is e^{-2} times that at the axis. As shown in Figure 5.3, the spatial coherence of the laser beam can be completely characterized by the minimum beam radius r_0 and the confocal parameter b, which is given by

$$b = \frac{2\pi r_0^2}{\lambda} \qquad (5.3)$$

At a longitudinal distance $z = b/2$, the beam area has expanded from its minimum value by a factor of 2. For $z > b/2$, beam divergence is approximately given by the far-field diffraction angle $\theta = \lambda/\pi r_0$. Typical values for a visible laser operating in the TEM_{00} mode are $\lambda = 500$ nm, $r_0 = 1$ mm, $b = 12$ m, and $\theta = 0.2$ mrad.

Clearly, between its high spectral purity and excellent spatial coherence, the laser would be the ideal source of radiation in photoacoustic experiments. Unfortunately, lasers to date usually operate at only a few discrete wavelengths or, if continuously tunable over a reasonably wide range, are still quite expensive and cumbersome. The most popular discrete line lasers in the visible optical region are the He–Ne laser, the ruby laser, the Nd:YAG laser, the argon ion laser, and the nitrogen laser.

The He–Ne laser is a CW (continuous-wave) gas laser that operates at 633 nm. An electral discharge excites the He atoms, which in turn excite the Ne atoms by means of radiationless collisions. The He–Ne laser is fairly inefficient and quite low in power (milliwatts).

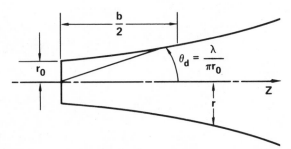

Figure 5.3 Parameters of a TEM_{00} Gaussian laser beam, with a minimum beam radius r_0 and a divergence θ_d.

The ruby laser was the first to exhibit optical lasing action. It also is relatively inefficient, requiring intense optical pumping to create a population inversion since the terminal laser level is the ground state. The long radiative lifetime of the upper level (3 msec), however, allows considerable energy storage during inversion and the ruby laser is usually used in a pulsed high-power mode at 694 nm.

The solid-state Nd:YAG laser radiates at 1.06 μm. The laser gain is large ($\sim$ 75 times that of the ruby laser), and at room temperature the population of the lower laser level is very low, thereby allowing threshold to be easily reached, both on a pulsed and CW basis. The Nd:YAG laser is a high-power and fairly efficient laser.

The argon ion laser operates on population inversions in doubly ionized, singly ionized, and atomic argon that are excited by electrical discharge. Highly stable CW emissions occur at several ultraviolet wavelength near 360 nm and at a dozen blue-green wavelengths in the 500-nm region. The argon ion laser is moderately efficient and is capable of producing 4 W in the ultraviolet and 20 W in the blue-green region of the visible spectrum.

The nitrogen laser radiates in the near ultraviolet at 337 nm in a pulsed mode. Because the upper laser level has a very short radiative lifetime (40 nsec), the emission at 337 nm is spread over 0.2 nm because the radiative buildup time is insufficient for the development of well-defined cavity modes (the photons will travel only 1.2 m in 40 nsec).

With the exception of the nitrogen laser, all the discrete line lasers described above have extremely high spectral purity and thus can be used for high-resolution photoacoustic spectroscopy. However, their narrow spectral ranges make them unsuitable for general spectroscopic studies where tunable radiation sources are required. All the above lasers, with the exception of the He–Ne laser, can be used to excite more broadly tunable sources of coherent visible and ultraviolet radiation, such as dye lasers, second-harmonic generators, optical parametric oscillators, frequency mixers, and third-order nonlinear optical generators.

By far the most promising source of tunable coherent radiation in the UV–visible region is the dye laser. With the use of selected dyes, this laser can be continuously tuned over the 340–1200 nm region with a spectral width of less than 1 nm (Deutsch, 1971). In addition, with the use of optical elements within the dye cavity, such as a diffraction grating or an etalon, linewidths of 10^{-6} Å are attainable for ultra-high-resolution work (Wu et al, 1974). In addition to its role as a primary source of continuously tunable narrow-band optical radiation, the dye laser can also be used to excite nonlinear devices that generate tunable coherent radiation as short as 100 nm. The tunability, spectral purity, and spatial coherence of dye lasers makes then an excellent source for high-resolution photoacoustic spectroscopy. However, dye lasers are still quite expensive, the dyes have short lifetimes, and tuning is often quite cumbersome.

5.6   COHERENT SOURCES: INFRARED

We define the infrared region to encompass 1–30 μm, a wavelength interval that contains most of the important molecular vibrational bands. We also distinguish between tunable and fixed-frequency lasers. A tunable laser has a total tuning range of several hundred reciprocal centimeters, although it may be continuously tunable over only 1 cm^{-1} at a time of this total range. A fixed-frequency laser may be step tunable from one line to another over a 100-cm^{-1} range, but is continuously tunable over only one narrow line (usually less than 0.01 cm^{-1}) at a time. For this reason fixed-frequency lasers have limited use in photoacoustic spectroscopy, since their lasing lines seldom match with the absorption bands of the molecules of interest, and sensitivity is therefore considerably reduced.

Broadly tunable lasers generally have some method of controlling the gross-tuning, as well as the fine-tuning, characteristics of the laser. This can be done by changing the gas in a gas laser, the composition in a semiconductor laser, or the dye in a dye laser. Fine tuning generally involves changes of a single cavity mode with an intracavity dispersive element, such as an etalon, or by small changes in the optical energy level spacing through changes in temperature or external electric or magnetic fields.

Fixed-frequency lasers tend to be more powerful than tunable lasers and thus are better suited to photoacoustic detection of small concentrations of gas. At atmospheric pressure and a 1-sec integration time a well-designed photoacoustic spectrometer can detect in the parts per million range with a milliwatt of average power, and in the parts per billion (ppb) range with a watt of average power for a typical infrared absorption cross section of 10^{-18} cm^2.

When high sensitivity is desired, high-powered fixed-frequency CO and CO_2 lasers have been used extensively. The CO laser has many lines in the 5–7 μm range (Mantz et al., 1970) while the CO_2 laser has a multitude of lines in the 9.2–11.5 μm region (Freed et al., 1974). Where high sensitivity is not of utmost importance, the broadly tunable lasers are more suitable infrared sources for photoacoustic spectroscopy. Table 5.1 gives some of the parameters for some of the more common tunable infrared laser systems.

Step tuning with nearly continuous coverage of many regions of the middle infrared can be obtained by sum- and difference-frequency genera-

TABLE 5.1 Some Tunable Infrared Lasers

Tunable source	Wavelength region (μm)	Typical power (W)	
		CW	Pulsed
Difference frequency generators	1–30	10^{-6} (2–4 μm)	10^5
Optical parametric oscillators	1–3.5 ($LiNbO_3$) 1.2–8.5 (Ag_3AsS_3) 8–12 (CdSe)	–	10^5
Semiconductor diode lasers	1–30	10^{-3}	10
Spin-flip Raman lasers	3 (HF pumped) 5–6 (CO pumped) 9–14 (CO_2 pumped)	1 (5μm)	10^3
High-pressure gas lasers	9–11 (CO_2)	–	10^5

Source: Kelley, P.L., 1977.

tion using the outputs of CO and CO_2 lasers. The sum and difference generation is performed with optical mixers and optical parametric oscillators. These devices involve nonlinear crystals in which a significantly large dielectric polarization can be induced that is quadratic in the strength of an applied laser field. Optical radiation can result from these nonlinear polarizations. Thus the second harmonics of the incident laser radiation can be generated at wavelengths of $\lambda/2$, where λ is the laser wavelength. In addition, if two laser beams are combined in a nonlinear crystal, sum- and difference-frequency generation occurs and radiation is emitted at $\lambda_1 \pm \lambda_2$ wavelengths, where λ_1 and λ_2 are the two laser wavelengths. The number of new wavelengths that can be generated by mixing is proportional to the product of the number of transitions of each laser. This can be substantially increased by including all the relatively abundant isotopic species, since the optical transitions are slightly shifted by changes in the electron–nucleus interaction.

With a tunable visible laser such as a dye laser, one can get broadly tunable infrared radiation through difference-frequency generation with either a ruby or argon ion laser.

The efficiency of sum- and difference-generation is fairly low. For example, by mixing 97 mW of CO radiation with 1.25 W of CO_2 radiation in a crystal of $CdGeAs_2$, only 4 μW of power was obtained at the difference wavelength of 13 μm (Kildal and Mikkelsen, 1974). Similarly, the genereration of 75 mW of CO_2 power at the second harmonic ($\lambda/2$) required 17 W of input power (Menyuk et al., 1976).

The optical parametric oscillator (OPO) consists of a nonlinear crystal within an optical cavity (Harris, 1969; Byer, 1975). A laser pumps the crystal at a frequency ω_p. Initially the pump radiation mixes with broadband photon noise, but because of the optical cavity there is a buildup of radiation at only two frequencies: the signal frequency ω_s and the idler frequency $\omega_i = \omega_p - \omega_s$. The signal frequency ω_s is that frequency from the photon noise that is resonant within the optical cavity. These OPO devices operate most efficiently with pulsed lasers.

In many photoacoustic studies of gaseous matter, the investigators have used a spin-flip Raman laser (Patel and Shaw, 1971). In this device, a fixed-frequency laser, such as a CO or CO_2 laser, pumps a semiconductor crystal (n-type InSb) in a magnetic field. Some of the pump-laser photons lose energy when they interact with an electron in the crystal and flip its spin. The emitted downshifted Raman photon is separated in energy from the pump photon by the magnetic spin energy $g\beta_e H$ of the electron, where g is the gyromagnetic ratio, β_e is the Bohr magneton, and H is the applied field. Thus by varying the applied field, the wavelength of the Raman photon can be continuously tuned over a small range. At sufficiently high pump power, stimulated emission of the Raman photons can exceed losses, and lasting action occurs. Fairly high conversion efficiency can be achieved by using a pump laser wavelength near the band gap of the semiconductor.

Molecular gas lasers, such as CO_2, have usually been operated at relatively low pressures of around 10 torr. Consequently, the gain bandwidths are essentially Doppler limited to approximately 50 MHz (~0.1 Å). Thus continuous tuning of this fixed-frequency laser is limited to 0.1 Å or 10^{-3} cm^{-1}. It is possible to increase the range of continuous tuning by increasing the natural linewidth of the transition through collisional broadening at high pressures. However, high-pressure discharges in a long laser tube are difficult to maintain, since in addition to the increased bandwidth, the deactivation rate is higher, leading to a higher pump threshold. Nevertheless, high-pressure gas lasers are under development and hold considerable promise. For example, continuous tuning over 5 cm^{-1} has been obtained in a high-pressure N_2O–CO_2 laser (Chang et al., 1976). In addition to their wider tunability, these high-pressure gas lasers should have higher peak and average power than any other tunable infrared source of radiation.

Perhaps the most promising new development in coherent infrared sources is the advent of recombination-radiation semiconductor lasers (Nathan, 1966; Melngailis and Mooradian, 1975). These lasers operate by stimulating emission across the gap between the conduction and valence bands of a semiconductor. Population inversion is achieved by electron injection across the band gap either by use of an electrical current (diode) or by optical pumping or electron-beam excitation. Infrared semiconductor laser materials in the 1–30 μm range include such binary compounds as InAs, InSb, GaSb, PbSe, PbS, and PbTe and such pseudobinary alloys as $Pb_{1-x}SnTe$, $PbS_{1-x}Se$, $Hg_{1-x}CdTe$, $InGa_{1-x}As$, and $GaAsSb_{1-x}$.

As is shown in Figure 5.4, coarse tuning of the infrared wavelength emitted by the lead salt diode lasers can be achieved by adjusting the chemical composition (the factor x in the formulas depicted). The composition determines the energy bandgap of the semiconductors and thus the wavelength of the spontaneous emission. Further tuning of a diode laser with fixed composition can be a accomplished by changing the temperature, applied pressure, or magnetic field that the semiconducting crystal sees. For ultrahigh-resolution spectroscopy, dispersive optical ele-

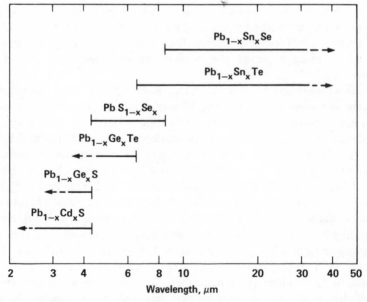

Figure 5.4 Compositional tuning ranges of lead salt diode lasers. (Reproduced by permission from Kelley, 1977.)

ments can be incorporated into the optical cavity formed by mirrors and the semiconducting laser.

At present these semiconductor lasers can be continuously tuned over only $0.1-1$ cm^{-1} by varying the operating temperature or drive current (Hinkley et al., 1968) or, alternatively, by varying an externally applied magnetic field (Melngailis and Mooradian, 1975). A broader tuning range is possible with hydrostatic pressure (Besson et al., 1968). With the broadband wavelength coverage available using hydrostatic pressure tuning, only the binary semiconductor compounds and one or two alloy semiconductors would be necessary to cover the entire wavelength range of $2-35$ μm.

A basic problem with these semiconductor lasers is their relatively low output power, providing only $1-20$ mW CW power in single-mode operation. Furthermore, the efficiency of these devices decreases rapidly if the operating temperature exceeds 77°K, because of a sharp increase in the threshold current necessary to substain lasing action. Once these devices are made to operate at room temperature and higher powers, they will probably become the optimal sources of infrared radiation for photoacoustic spectroscopy.

REFERENCES

Besson, J. M., Paul, W., and Calawa, A. R. (1968). *Phys. Rev.* **173**, 699.

Billings, B. H. (1972). In *American Institute of Physics Handbook*, 3rd ed., pp. 6-216–6-217. McGraw-Hill, New York.

Born, M. and Wolf, E. (1965). *Principles of Optics*, Pergamon, Oxford.

Byer, R. L. (1975). In *Quantum Electronics*, Vol. 1B, (H. Rabin and C. L. Tang, Eds.), p. 587, Academic Press, New York.

Calvert, J. G., and Pitts, J. N., Jr., (1966). *Photochemistry*, Wiley, New York.

Cann, M. W. P. (1969). *Appl. Opt.* **8**, 1645.

Chang, T. Y., McGee, J. D., and Wood, O. R. (1976). *Opt. Commun* **18**, 279.

Deutsch, T. F. (1971). In *Handbook of Lasers* (R. J. Pressley, Ed.), pp. 350–354, Chemical Rubber Co., Cleveland, Ohio.

Freed, C., Ross, A. H. M., and O'Donnell, R. G. (1974). *J. Mol. Spectrosc.* **49**, 439.

Hackforth, N. L. (1960). *Infrared Radiation*, McGraw-Hill, New York.

Harris, S. E. (1969). *Proc. IEEE*, **57**, 2096.

Hinkley, E. D., Harman, T. C., and Freed, C. (1968). *Appl. Phys. Lett.*, **13**, 49.

Jacquinot, P. (1960). *Rep. Prog. Phys.* **23**, 267.

Kasper, J. (1972). In *American Institute of Physics Handbook*, 3rd ed., p. 6-198, McGraw-Hill, New York.

Kelley, P. L. (1977). In *Optoacoustic Spectroscopy and Detection* (Y.-H. Pao, Ed.), p. 122, Academic Press, New York.

Kildal, H., and Mikkelsen, J. C. (1974). *Opt. Commun.* **10**, 306.

Koller, L. R. (1965). *Ultraviolet Radiation*, 2nd ed., Wiley, New York.

Mantz, A. W., Nichols, E. R., Alpert, B. D., and Rao, K. N. (1970). *J. Mol. Spectrosc.* **35**, 325.

McNesby, J. R., Braun, W., and Ball, J. (1971). In *Creation and Detection of the Excited State* Vol. 1, Part B. (A. A. Lamola, Ed.) pp. 503–582, Dekker, New York.

Melngailis, I., and Mooradian, A. (1975). In *Laser Applications to Optics and Spectroscopy* (S. F. Jacobs, M. Sargent, J. F. Scott, and M. O. Scully, Eds.), p. 1. Addison-Wesley, Reading, Mass.

Menyuk, N., Isler, G. W., and Mooradian, A. (1976). *Appl. Phys. Lett.* **29**, 422.

Nathan, M. I. (1966). *Proc. IEEE* **54**, 1276.

Patel, C. K. N., and Shaw, E. D. (1971). *Phys. Rev.* **B3**, 1279.

Samson, J. A. R. (1967). *Techniques for Vacuum Ultraviolet Spectroscopy*, Wiley, New York.

Tanaka, Y. (1955). *J. Opt. Soc. Am.* **45**, 710.

Wolfe, W. L. (Ed.), (1965). *Handbook of Military Infrared Technology* Naval Research Dept., U.S. Govt. Printing Office, Washington.

Wu, F. Y., Grove, R. E., and Ezekiel, S. (1974). *Appl. Phys. Lett.* **25**, 73.

6

PHOTOACOUSTIC
SPECTROSCOPY OF GASES

6.1 INTRODUCTION

As we indicate in earlier chapters, the photoacoustic method has been used fairly extensively in gas analysis and vapor detection since the original descriptions of these applications by Viengerov (1938) and by Luft (1943). In this chapter we discuss the more recent work in this field.

6.2 SIGNAL STRENGTH

The theory of the photoacoustic effect in gases (Chapter 3) shows that the signal strength is directly proportional to the amount of power absorbed. This in turn is related to the intensity and monochromaticity of the incident light and to the concentration of the absorbing species in the photoacoustic cell. The fact that the principal sources of noise in a photoacoustic system are acoustic and electronic results in the signal/noise ratio increasing linearly with power absorbed, which provides a considerable advantage over conventional optical absorption measurements. For example, let us consider the detection of CO_2 with a conventional blackbody IR source operating at 800°K. With such a source it is possible to obtain an average power of 5 μW in a 0.04-μm bandwidth around 4.283 μm, which corresponds to a fundamental absorption band of CO_2. Typical sensitivities of commercial absorption detectors would then be of the order of parts per million (Hill and Powell, 1968). On the other hand, a photoacoustic system employing a resonant cell would have a sensitivity of $\sim$0.1 ppm with the same source (Dewey, 1974).

The high power available from lasers permits even greater sensitivity of photoacoustic detection (ppb). In addition, the narrow bandwidth of laser sources gives improved selectivity over blackbody sources. The use of a laser as an exciting source in gas photoacoustic systems was first reported by Kerr and Attwood (1968) and has subsequently been described by numerous other authors (Kreuzer, 1971; Kreuzer and Patel, 1971; Kreuzer et al., 1972; Patel et al., 1974; Dewey et al., 1973; Goldan and Goto, 1974; Claspy et al., 1976).

There are disadvantages as well as advantages to using lasers as sources. The primary disadvantage is that no broad-band tunable lasers exist in the infrared, and thus high sensitivity vapor detection is limited to those compounds whose absorption bands overlap the wavelength regions of the present infrared lasers. Nevertheless, the high sensitivity and linearity of response with concentration of laser-excited photoacoustic systems make them useful tools for detecting and monitoring atmospheric pollutants and for conducting measurements of concentration of infrared-absorbing molecules in the atmosphere. In addition, photoacoustic systems provide a capability for detection of the presence of illegal materials with low vapor pressure, such as explosives and drugs. In all applications, the investigator must also deal with the presence of many other infrared-absorbing species in the atmosphere. Thus in designing a system, consideration must be given both to the absorption spectrum of the species to be studied and to the absorption spectrum of any anticipated interfering species.

6.3 TYPICAL SYSTEM

A typical laser-driven photoacoustic spectrometer for gas studies is shown schematically in Figure 6.1. The laser shown is a CO or CO_2 laser, whose

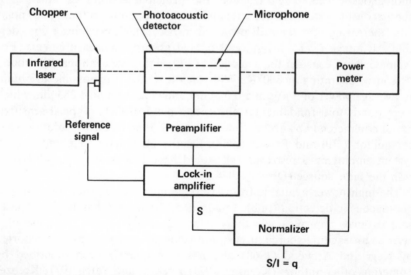

Figure 6.1 A schematic diagram of a typical gas photoacoustic spectrometer. (Reproduced by permission from Claspy, 1977.)

intensity is modulated by an electromechanical chopper at some acoustic frequency appropriate for the photoacoustic cell. The cell may be one of the several designs that we discuss in Chapter 4. The power meter is the reference detector, used to monitor the optical power in the laser beam. Since we are usually dealing with the detection of very low concentrations, the power meter reads the power of the beam exiting from the photo-acoustic cell, which is essentially identical to the power entering the cell. The power meter, or some other reference detector, is necessary to normal-ize the variations in the incident laser power arising from changes in wavelength and operating conditions. The photoacoustic cell contains a condenser microphone that is either an externally biased or an electret type. The signal from the microphone is processed by a suitable preampli-fier and a phase-sensitive lock-in amplifier tuned to the modulation frequency of the light chopper.

6.4   THE MICROPHONE

The detector used in gas photoacoustic studies is a condenser microphone of either the electret type or the externally biased type. Microphones are displacement-sensitive devices, and, as such, they are particularly sensitive to the temperature-induced pressure–volume changes in a photoacoustic cell. Condenser microphones consist of a thin metal diaphragm or metal-lized plastic dielectric and a rigid conducting back plate. If a charge is applied to this capacitor by an external d.c. supply in an electrically biased microphone, capacitance modulation caused by sound–pressure induced changes in the plate separation produce a current flow between the plates. The signal output from the microphone therefore depends on the micro-phone capacitance, the pressure-induced changes in capacitance, and the magnitude of the applied voltage. The signal is also inversely proportional to the total capacitance of the circuit, so it is essential that the microphone capacitance, the amplifier input capacitance, and any stray capacitance be kept to a minimum.

An equivalent circuit diagram of a condenser microphone is shown in Figure 6.2 (*Microphones and Microphone Preamplifiers*, 1975). The output voltage is

$$V_0(t) = \frac{\Delta C(t)}{C_t} V_B \frac{i\omega RC_t}{1 + i\omega RC} \tag{6.1}$$

where V_B is the bias voltage, $C(t)$ is the variation in capacitance due to sound–pressure fluctuations, and C is the total capacitance given by

$$C = C_t + C_s + C_i \tag{6.2}$$

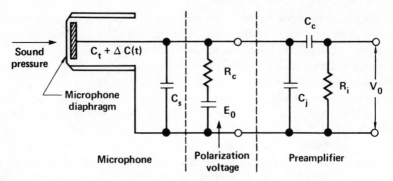

Figure 6.2 Equivalent circuit diagram of a condenser microphone and preamplifier. (Reproduced by permission from *Microphones and Microphone Amplifiers*, 1975.)

where C_t is the microphone capacitance, C_s is the stray capacitance, and C_i is the preamplifier input capacitance. The resistance R is given by

$$R = \frac{R_i R_c}{R_i + R_c} \tag{6.3}$$

where R_i is the preamplifier input resistance and R_c is the charging resistance.

The microphone sensitivity is defined as

$$S = \frac{V_0(t)}{\Delta p(t)} \tag{6.4}$$

where $\Delta p(t)$ is the sound–pressure variation with time. Two basic cases can be considered in terms of frequency response:

1. $\omega RC \gg 1$

$$S \simeq S_0 V_B \left(\frac{C_t}{C} \right) \tag{6.5}$$

where $S_0 = \Delta C(t)/C_t \Delta p(t)$ is the open-circuit sensitivity. Thus at high frequencies, the sensitivity is reduced by the stray capacitance of the cables and the input capacitance of the preamplifier. For this reason, the pre-amplifier is usually located as close as possible to the microphone.

2. $\omega RC \ll 1$

$$S \simeq S_0 V_B i \omega R C_t \tag{6.6}$$

At low frequencies then, the sensitivity decreases with decreasing frequency. To obtain low-frequency sensitivity, the preamplifier input resistance must be large ($> 10^8$ Ω). Such high input impedances are available in FET amplifiers. Typical sensitivities of good microphones are 1–5 mV per μbar (1 dyne/cm^2) of pressure.

6.5 WAVELENGTH SELECTION

The wavelengths to be used for performing the measurement depend on both the infrared absorption spectrum of the gas to be studied and on the spectra of any other infrared-absorbing species that may be present. Outside the homonuclear diatomic molecules, nearly all gases absorb in the 2–15 μm region. In this range, only the CO and CO_2 lasers are practical sources for photoacoustic systems. While sources such as tunable infrared diode lasers and spin-flip Raman lasers have been used to some extent, most PAS measurements are confined to the 5–7 μm range of the CW CO laser and the 9.2–11.5 μm range of the CW CO_2 laser. Although neither the CO laser or the CO_2 laser is continuously tunable over its respective range, both lasers have a multitude of lines, with average line separations being no more than 2–4 cm^{-1}. This results in frequent near or actual coincidences between the laser emission lines and the gas absorption lines. The discrete tunability of the CO and CO_2 laser is, in fact, an advantage in the detection of small-molecule vapors, which tend to have narrow absorption lines. The high degree of wavelength stability and reproducibility of gas lasers (Freed, 1970, 1971) makes possible accurate and reproducible analysis of small-molecule vapors.

6.6 SMALL-MOLECULE GASES

The absorption spectra of gases of low molecular weight are characterized by sharp spectral lines that remain fairly sharp even when pressure broadened at 1 atm.

Figure 6.3 shows the absorption spectrum of a hypothetical small-molecule gas in the region of the CO_2 laser emission region. The solid curve consists of several pressure-broadened lines, the tails of which are represented by the dashed lines. The vertical lines starting at the x-axis represent the CO_2 laser emission lines. Because of the discrete natures of both spectra, there is only one near coincidence, indicated by the point P.

Unless the composition of the air sample is known precisely, it is usually not advisable to rely soley on a single coincidence, since it is not unusual

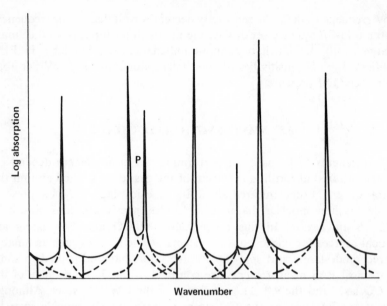

Figure 6.3 Sampling of small-molecule gas absorption at CO_2 laser emission wavelengths. (Reproduced by permission from Kreuzer, 1974.)

for several similar gases to have one or more nearly identical coincidences. However, if the PAS signal is recorded at at least four or five laser wavelengths, the pattern of PAS signals is unique for each molecule, and identification errors are considerably reduced. In practice, it is usual to record the PAS signal over the entire tuning range of the laser to minimize identification errors.

There have been several recent papers on the detection of atmospheric pollutants. These include a study of NO concentrations resulting from highway traffic using a spin-flip Raman laser as the source (Kreuzer and Patel, 1971). The PAS spectra taken from this study are shown in Figure 6.4. Figure 6.4a shows the calibration spectrum of 20 ppm NO in N_2 in the region from 1813 to 1823 cm^{-1}. The spectrum consists of five NO lines numbered 1, 5, 6, 8, and 11, while the remaining lines are due to H_2O. Figure 6.4c is the spectrum of laboratory air at 21°C and 30% relative humidity, indicating the strong water vapor absorption at either end of the spectral range. The analysis in Kreuzer and Patel's paper was thus based on the NO lines 6 and 8, which lie in the central region of the range. Figures 6.4d and 6.4e are spectra taken near the highway and from an automobile exhaust, respectively, and show that a sensitivity of ~2 ppm

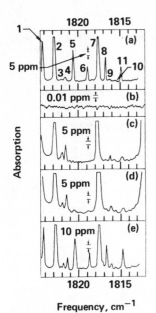

Figure 6.4 Absorption spectra of various gas samples in the range of 1815–1825 cm^{-1} obtained by changing the magnetic field (of a spin-flip Raman laser) from 30 to 40 kG. All spectra are taken at a pressure of 500 torr. (*a*) Calibration spectrum of 20 ppm NO in N_2. Lines numbered 1, 5, 6, 8, and 11 are due to NO; the remaining lines are due to H_2O. Vertical sensitivity is 1.0 and integration time is 4 sec. (*c*) Room air is at 21°C and 30% relative humidity, vertical sensitivity is 1, and integration time is 1 sec. (*d*) Air sample from Route 22 near Plainfield, NJ. Vertical sensitivity is 1 and integration time is 1 sec. (*e*) Automobile exhaust. Vertical sensitivity is 0.4 and integration time is 1 sec. (Reproduced by permission from Kreuzer and Patel, 1971. Copyright 1971 by the American Association for the Advancement of Science.)

TABLE 6.1. Noise-Limited Sensitivities for Detecting Pollutant Gases

Gas	Sensitivity (ppb)	Infrared Source Laser	Transition	Wavelength (μm)
Ammonia	0.4	CO	$P_{19\text{-}18}(15)$	6.1493
Benzene	3.0	CO_2	$00.°1\text{–}02°0\ P(30)$	9.6392
1,3-Butadiene	1.0	CO	$P_{20\text{-}19}(13)$	6.2153
1,3-Butadiene	2.0	CO_2	$00°1\text{–}10°0\ P(30)$	10.6964
1-Butene	2.0	CO	$P_{19\text{-}18}(9)$	6.0685
1-Butene	2.0	CO_2	$00°1\text{–}10°0\ P(38)$	10.7874
Ethylene	0.2	CO_2	$00°1\text{–}10°0\ P(14)$	10.5321
Methanol	0.3	CO_2	$00°1\text{–}02°0\ P(34)$	9.6760
Nitric oxide	0.4	CO	$P_{8\text{-}7}(11)$	5.2148
Nitrogen dioxide	0.1	CO	$P_{20\text{-}19}(14)$	6.2293
Propylene	3.0	CO	$P_{19\text{-}18}(9)$	6.0685
Trichloroethylene	0.7	CO_2	$00°1\text{–}10°0\ P(24)$	10.6321
Water	14.0	CO	$P_{17\text{-}16}(13)$	5.9417

Source: Kreuzer et al., 1972. Copyright 1972 by the American Association for the Advancement of Science.

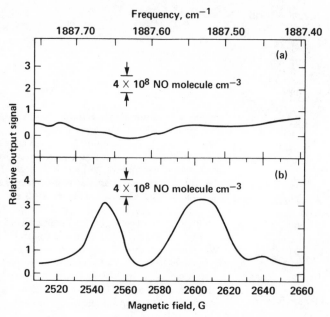

Figure 6.5 (a) Relative output signal as a function of the spin-flip Raman (SFR) laser magnetic field for NO, taken before ultraviolet sunrise (11:55:00–12:27:00 G.m.t., October 19, 1973). The NO absorption signals are expected to occur at magnetic fields of 2545 and 2605g. Calibration in equivalent NO molecules per cubic centimeter is given. (b) Relative output signal as a function of the SFR laser magnetic field for NO taken after ultraviolet sunrise (17:21:50–17:52:05 G.m.t. October 19, 1973). NO absorption signals at magnetic fields 2545 and 2605g can be seen. Calibration in equivalent NO molecules per cubic centimeter is given, indicating a NO concentration of $(2 \pm 0.15) \times 10^9$ molecule/cm^3. (Data taken at 28-km altitude and 225°K temperature.) (Reproduced by permission from Patel et al., 1974. Copyright 1974 by the American Association for the Advancement of Science.)

NO can be obtained from the dry nitrogen, but a sensitivity of only 50 ppm NO is achievable in the presence of water vapor in normal air. Kreuzer et al. (1972) have measured minimum sensitivities for detection of several atmospheric pollutants using a nonresonant photoacoustic cell. The results of these measurements are shown in Table 6.1. In this study, CO and CO_2 lasers were used.

Patel et al. (1974) measured the concentration of NO as a function of altitude with a balloon-borne PAS system. In this experiment two No absorptions near 1887 cm^{-1} were selected for identification purposes. Figures 6.5a and 6.5b are spectra of air samples taken at an altitude of 28 km before and after ultraviolet sunrise, respectively. Figure 6.5a shows no structure and therefore indicates that the NO concentration was less than 1.5×10^8 cm^{-3} in the absence of sunlight. Figure 6.5b was taken 5.5 hours after sunrise, at which time the spectrum showed two distinct NO absorption bands indicating an NO concentration of $\sim 2 \times 10^{19}$ cm^{-3}.

6.7 LARGE-MOLECULE GASES

The approach used to detect and identify large-molecule gases and vapors is somewhat different than that used for small-molecule gases. As we note earlier, lighter-molecule gases exhibit sharp absorption lines. Large-molecule gases tend to exhibit unresolved continuum spectra several wavenumbers in width even at low pressures.

In PAS experiments on large-molecule gases one must locate a band spectral feature characteristic of the molecule and map it out as thoroughly as possible to prevent identification errors. For example, the -NO$_2$ group, which occurs in most explosives, exhibits an asymmetric stretch-vibrational absorption band near 6 μm and an -O-N stretching mode near 11 μm, with the specific location being determined by the basis molecule to which the group is attached. Peak asymmetric stretch bands of -NO$_2$ occur at 6.039 and 6.086 μm in ethylene glycol dinitrate (EGDN) and at 5.970, 6.046, and 6.079 μm for nitroglycerine (NG). The peak -O-N absorptions occur at 11.148, 11.628, and 11.877 μm for EGDN and at 11.099 and 11.905 μm for NG (Pristera et al., 1960).

While the use of spectral characteristics such as the ones described above make detection and identification possible, the effects of interference from other large-molecule species and from water vapor can be quite significant and considerably more serious than in the case of small-molecule gases.

The presence of water vapor is particularly acute in the 5–7 μm region. Gelbwachs (1974) estimated that for the nearly 4000 water vapor absorption lines in this region, fewer that 20% of the spectral intervals between lines exceed 9 GHz or 0.3 cm^{-1}. He also calculated that the limiting sensitivity for detection at 1 atm is $\sim 10^3$ and 0.1 ppb in the proximity of strong and weak water vapor lines, respectively.

Claspy et al. (1976) performed a feasibility study for the detection of large-molecule gases such as nitroglycerine, dinitrotoluene (DNT), and ethylene glycol dinitrate. In this experiment the PAS signal was obtained at four different laser lines in three wavelength regions, near 6, 9.4, and 11 μm. Although the minimum detectable concentration per milliwatt of laser

TABLE 6.2. **Minimum Detectable Explosive Concentration Using Four Wavelengths**

Wavelength (μm)	Explosive	Minimum Detectable Concentration (ppm vapor)
6	NG	0.24×10^1
	EGDN	2.5
	DNT	220.0
9	NG	0.055×10^{-2}
	EGDN	2.5
	DNT	16.0
11	NG	0.028×10^{-2}
	EGDN	1.5

Source: Claspy et al., 1976.

power is nearly the same in each spectral region when dry N_2 is used as the carrier gas, as shown in Table 6.2, the situation becomes significantly more difficult at 6 μm when air with a 50% relative humidity is used. In this case the water vapor signal masks the explosive vapor signal so effectively that detection here is virtually impossible. Fortunately, detection at 9.4 and 11 μm is unaffected by the water vapor.

6.8 MULTICOMPONENT SAMPLES

In gas PAS studies it is often desirable to analyze a sample for a number of different species, rather than just one. This can be done by measuring the PAS signal at each of a set of wavelengths chosen on the basis of the absorption spectra of the individual components. If the response of a PAS system at wavelength λ_i per unit incident power is designated as R_i, then the signal S_i is given by

$$S_i = P_i c R_i \qquad (6.7)$$

where c is the concentration and P_i is the power at λ_i. For a multicomponent sample

$$S_i = P_i \sum_{n=1}^{N} R_{in} c_n \qquad (6.8)$$

where R_{in} is the response due to component n at i. Formally, the solution

TABLE 6.3. Rejection Ratios

Component Being Measured	Interfering Component			
	Ethanol	Methanol	Ammonia	Trichloroethylene
Ethanol	—	270	3200	16,000
Methanol	760	—	1900	300
Ammonia	1080	430	—	1,080
Trichloroethylene	200	200	1000	—

Source: Kreuzer, 1974.

to (6.8) is

$$c_n = \sum_{i=1}^{N} R_{ni}^{-1}\left(\frac{S_i}{P_i}\right) \tag{6.9}$$

where R_{ni}^{-1} is the inverse of the matrix R_{in}.

The effectiveness of this method in analyzing a multicomponent sample depends not only on the precision of the measurement of S_i and P_i, but also on the nature of the matrix R_{in}. Clearly, if a set of wavelengths can be selected such that R_{in} is diagonal, that is, if at each wavelength we get a response from only one component, then our problem is trivial and the sensitivity is not limited by interference among the components.

Such a happy state is generally impossible to achieve, and usually considerable attention must be given to the ability of the system to discriminate among the various known components of the mixture, as well as to spurious results arising from the presence of any unexpected components.

The effect of a large concentration of one gas A on the ability of a system to detect a small concentration of a second gas B is denoted by a rejection ratio. This parameter is defined as the concentration of gas A necessary to give the same signal as a unit concentration of gas B at the wavelength chosen for gas B. Table 6.3 [from Kreuzer (1974)] gives the rejection ratio for a number of molecules.

6.9 OPTICAL SATURATION

We state at the beginning of this chapter that the photoacoustic signal increases directly as the power of the incident beam increases. It would thus appear that the sensitivity of gas PAS systems could be improved indefinitely by simply going to ever more powerful lasers.

Unfortunately, this is not the case, since at sufficiently high intensities, optical saturation effects come into play. As we show in Chapter 3, the photoacoustic signal in a gas is given by

$$q = \frac{kE_1N^2}{C_v\omega}\left\{\frac{2\tau_c^{-2}BI_0\delta}{\left[2BI_0+\tau^{-1}\right]\left[\left(2BI_0+\tau^{-1}\right)^2+\omega^2\right]^{1/2}}\right\}e^{i(\omega t-\gamma+\pi/2)} \quad (6.10)$$

where k is the Boltzmann constant, N is the number of gas molecules per cm^3, E_1 is the energy of the excited level, C_v is the specific heat, ω is the modulation frequency, τ is the collisional lifetime, B is the Einstein coefficient, I_0 is the beam intensity, δ is the modulation function, τ is the total lifetime of the upper level, and γ is the phase delay due to the deexcitation processes ($\gamma = \omega\tau$).

For low intensities ($I_0 < 1/B\tau$), (6.10) becomes

$$q \simeq \frac{kE_1N^2}{C_v\omega}\left(\frac{\tau}{\tau_c}\right)^2\frac{2BI_0\delta}{(1+\omega^2\tau^2)^{1/2}}e^{i(\omega t-\gamma+\pi/2)} \quad (6.11)$$

and thus q is linearly proportional to I_0.

At high intensities ($I_0 > 1/B\tau$),

$$q \simeq \frac{kE_1N^2}{C_v\omega}\tau_c^{-2}\frac{\delta}{BI_0}e^{i(\omega t-\gamma+\pi/2)} \quad (6.12)$$

Thus at high intensities q ceases to increase with increasing I_0 and in fact begins to decrease as $1/I_0$ at very high intensities. This condition is known as optical saturation.

Since τ is usually of the order of 10^{-6} to 10^{-4} sec, saturation starts to set in for most gases at W/cm^2 intensity. It is thus clearly important to avoid using unnecessarily high incident powers, since the signal will begin to decrease, rather than increase. Pao and Claspy (1975) have reported optical saturation in a photoacoustic experiment with nitroglycerine when I_0 approached 1 W/cm^2.

Thus higher detectability with photoacoustic spectroscopy must be achieved by reduction in the noise level, rather than by simply using ever more powerful laser sources.

REFERENCES

Claspy, P. C. (1977). In *Optoacoustic Spectroscopy and Detection* (Y.-H. Pao, Ed.), pp. 133–166, Academic Press, New York.

Claspy, P. C., Pao, Y.-H., Kwong, S., and Nodov, E. (1976). *Appl. Opt.* **15**, 1506.

Dewey, C. F., Jr. (1974). *Opt. Eng.* **13**, 483.

Dewey, C. F., Jr. Kamm, R. D., and Hackett, C. E. (1973). *Appl. Phys. Lett.* **23**, 633.

Freed, C. (1970). *Int. Electron Devices Meet.*, Washington, D. C. Paper 14.4.

Freed, C. (1971). *Appl. Phys. Lett.*, **18**, 458.

Gelbwachs, J. (1974). *Appl. Opt.* **13**, 1005.

Goldan, P. D., and Goto, K. (1974). *J. Appl. Phys.* **45**, 4350.

Hill, D. W., and Powell, T. (1968). *Non-Dispersive Infra-Red Gas Analysis in Science, Medicine and Industry*, Plenum Press, New York.

Kerr, E. L., and Attwood, J. G. (1968). *Appl. Opt.* **7**, 915.

Kreuzer, L. B. (1971). *J. Appl. Phys.* **42**, 2934.

Kreuzer, L. B. (1974). *Anal. Chem.* **46**, 235A.

Kreuzer, L. B., and Patel, C. K. N. (1971). *Science* **173**, 45.

Kreuzer, L. B., Kenyon, N. D., and Patel, C. K. N. (1972). *Science* **177**, 347.

Luft, K. F. (1943). *Z. Tech. Phys.* **24**, 97.

Microphones and Microphone Preamplifiers (1975). Bruel and Kjair, DK-2850, Naerum, Denmark.

Pao, Y.-H., and Clapsy, P. C. (1975). "An Investigation of the Feasibility of Use of Laser Optoacoustic Detection for the Detection of Explosives," Case Western Reserve University, Final Report for Subcontract 44343-V, The Aerospace Corp.

Patel, C. K. N., Burkhardt, E. G., and Lambert, C. A. (1974). *Science* **184**, 1173.

Pristera, F., Halik, M., Castelli, A., and Fredericks, W. (1960). *Anal. Chem.* **32**, 495.

Viengerov, M. L. (1938). *Dokl. Akad. Nauk. SSSR* **19**, 687.

CHAPTER

7

DEEXCITATION STUDIES IN GASES

7.1 INTRODUCTION

The possibility of using the photoacoustic effect to investigate deexcitation processes in gases was realized many years ago. In 1946 Gorelik proposed that measurement of the phase of the photoacoustic signal in a gas system could be used to study the rate of energy transfer between the vibrational and the translational degrees of freedom of the gas molecules. When a sample gas in a photoacoustic cell is irradiated by photons, which it absorbs, the absorbed energy excites a vibrational or vibrational–rotational energy state if the irradiation is in the infrared, or the absorbed energy excites an electronic energy level if the irradiation is in the visible or ultraviolet region of the spectrum. After a time delay determined by the rate of energy transfer through interatomic collisions, the excited state deexcites with a transfer of energy to the translational modes of the gas molecules, causing the gas to heat up through an increase in the kinetic energy of the gas molecules. If the irradiation time is long compared to the time required for this energy transfer, then essentially all the absorbed energy appears as heat energy, and the resultant photoacoustic signal is in phase with the incident radiation. If the frequency at which the incident light is chopped or modulated is high enough such that the irradiation time is less than the time required for energy transfer to occur, then not all the absorbed energy appears as periodic heat, and the phase of the maximum pressure fluctuation is noticeably different than that of the incident radiation. Thus a study of the phase of the photoacoustic signal as a function of modulation frequency gives information about the rate of intermolecular energy transfer. Gorelik's proposal was successfully put into practice for the first time by Slobodskaya in 1948.

7.2 DEEXCITATION DYNAMICS

The original deexcitation studies were performed in the infrared, where usually a low-level vibrational state was excited and then decayed to the ground state without populating intermediate levels. This situation can be treated as a simple two-level system as shown by Delaney (1959) and Read (1967–1968).

We can consider the thermodynamic state of the gas to be represented by two temperatures, T_1, corresponding to the vibrational modes, and T_2, corresponding to the translational modes. If the molecules in each temperature group are in statistical equilibrium within the group, then the energy exchange between these two groups can be characterized by a relaxation time τ. Thus

$$\frac{d\,\Delta T}{dt} = -\frac{\Delta T}{\tau} \tag{7.1}$$

where

$$\Delta T = T_1 - T_2 \tag{7.2}$$

We then find that for harmonic irradiation at frequency ω

$$\Delta T = \Delta T_{\max} \cos(\omega t - \phi) \tag{7.3}$$

where

$$\Delta T_{\max} = \frac{A}{C_v \omega (1 + \omega^2 \tau^2)^{1/2}} \tag{7.4}$$

and

$$\phi = \frac{\pi}{2} - \tan^{-1} \omega \tau \tag{7.5}$$

Here A is a constant relating to the product of the light intensity and the oscillator strength of the transition in question and C_v is the heat capacity of the gas. The photoacoustic signal itself is proportional to ΔT.

In the case of excitation to electronic energy levels, an initial excitation corresponding energetically to many quanta of vibration is usually involved, and the pathway to heat is far less certain. Hence the determination of a lifetime, as measured in the infrared, is less clear for an electronic transition. As stressed by Cottrell et al. (1966), the relaxation time measured photoacoustically is not necessarily that appropriate to the level initially excited, but may correspond instead to the average lifetime of energy being locked up in a variety of modes before becoming heat and activating the microphone. This complication is especially important for electronic excitations in large molecules, where the intermediate levels may be either vibrations or low-lying electronic states of the molecule. It may happen as well that the intermediate levels are within other molecules that are excited by intersystem crossings.

Upon excitation into an electronically excited state, a molecule can deexcite in any of three modes:

1. It may deexcite radiatively, through fluorescence or phosphorescence, to a vibrational level of the ground state, which then relaxes further by a nonradiative, or heat-producing, process.

2. It may undergo photochemistry to produce new chemical species.

3. It may deexcite nonradiatively through various intermediate electronic and vibrational energy levels, giving up all the initially absorbed energy into heat.

A relaxation scheme for a typical electronic excitation in an organic molecule has been constructed by Hunter et al. (1974) and is shown in Figure 7.1. They consider optical excitation from S_0 to S_1 followed by simultaneous return to S_0 by means of both fluorescence (releasing heat E_1, with rate constant k_1) and nonradiative transitions (releasing heat E_2, with rate constant k_2). In addition, they also include the possibility of an intersystem crossing to the lowest triplet T_1 (releasing heat E_3, with rate constant k_3). Those molecules reaching T_1 then relax radiationlessly to release heat E_4 at rate k_4. Hunter et al. show that for unimolecular decay at each step, the phase angle ϕ is given by

$$\tan\phi = L(\sin\theta_T\cos\theta_T)(H + L\cos^2\theta_T) \qquad (7.6)$$

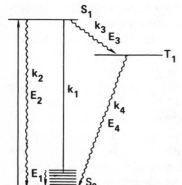

Figure 7.1 Relaxation scheme for a typical electronic excitation in an organic molecule (Reproduced by permission from Hunter et al., 1974.)

where

$$\cos \theta_T = \left(1 + \omega^2 \tau_4^2\right)^{-1/2} \qquad (7.7)$$

provided the lifetimes in the singlet manifold are very much smaller than either τ_4 or ω^{-1}, which is almost always the case. In (7.6), L represents the amount of heat resulting from relaxation in the slow T_1 step and H is the sum of the heats that are faster and either precede or parallel the excitation of T_1. Similarly, the amplitude of the photoacoustic signal is given by

$$q = \left(\frac{bF_0}{C_v \omega}\right)\left[H^2 + \cos^2 \theta_T (L^2 + 2HL)\right]^{1/2} \qquad (7.8)$$

in which bF_0 is a measure of the absorbed power per unit volume.

If $L \gg H$, as is the case when E_3 is small and k_3 is very large so that virtually all the molecules pass through T_1, then (7.6) and (7.8) reduce to (7.5) and (7.4), as given in the simpler two-level thermodynamic treatment.

Measurement of the phase angle at any one wavelength for a polyatomic molecule yields a single value for the lifetime, even though there are in fact several consecutive and/or simultaneous deexcitation steps with different lifetimes. We can consider each relaxation step as a vector, with a magnitude proportional to the amount of heat released, and the tangent of its angle depending on the lifetime. In a complex system then, we can have a nonunique relaxation pathway consisting of several successive steps, each step characterized by a relaxation time τ_i. The phase angle due to this one pathway is then given by

$$\tan \phi_j = \sum_i \tan \phi_i \qquad (7.9)$$

If there are several parallel pathways to the ground state, each with its overall phase angle ϕ_j and heat amplitude A_j, the overall heat vector is then the vector sum of all the individual vectors, and the measured phase is given by

$$\tan \phi_T = \frac{\sum_j A_j \sin \phi_j}{\sum_j A_j \cos \phi_j} \qquad (7.10)$$

Unless there is one particular level i that dominates in terms of magnitude or time, the measured lifetime has no meaning in terms of the lifetime of a particular level. In this regard it is more meaningful to obtain a time analysis of the heat pulse shape arising from a pulse of optical excitation.

In studying deexcitation phenomena in gas photoacoustic systems one must be careful to stay away from any acoustic resonances of the cell, including Helmholtz resonances, since a resonance grossly disturbs the phase measurement. Also, one must keep in mind that heat released by the gas molecules can diffuse to the cell walls and introduce a frequency-dependent phase lead while decreasing the amplitude of the signal. This effect can become quite large at low pressures (below 1 torr) and for chopping frequencies that are not much greater than the reciprocal of the characteristic time for heat conduction to the wall. Finally, infrared studies have shown that impurities at the parts per million level can be very active in promoting vibrational relaxation, and this should be considered equally important for electronic level relaxation, particularly in the case of oxygen impurities.

Below we illustrate some of the deexcitation experiments performed on gaseous samples with examples of photochemical processes, intersystem crossing, and impurity quenching. Most of these examples are taken from the excellent series of experiments performed by Robin and his co-workers at Bell Laboratories.

7.3   PHOTOCHEMICAL DEEXCITATION

A good example of a photoacoustic study of a photochemical process is shown in Figure 7.2 (Harshbarger and Robin, 1973a), which includes both the optical absorption and the photoacoustic spectrum of NO_2. Although taken with far lower resolution, the photoacoustic spectrum follows the general features of the absorption spectrum up to about 4000 Å. Here the absorption peaks and then begins a slow decrease, while the photoacoustic spectrum exhibits a sharp drop of almost 30%. It is at 4000 Å that the dissociation limit of NO_2 is reached:

$$NO_2 \rightarrow NO(^2\Pi) + O(^3P) \qquad (7.11)$$

Under the relatively high-pressure conditions (10 torr) of the experiment, there is no fluorescence and thus the drop in photoacoustic signal cannot be attributed to that. The recombination of NO and O is rather slow, but the O atom can go on to react exothermically with another NO_2 molecule

$$O + NO_2 \rightarrow NO + O_2 + 47 \, kcal/mole \qquad (7.12)$$

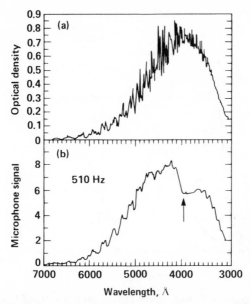

Figure 7.2 Comparison of the (a) optical and (b) photoacoustic spectra of NO_2 at 10 torr. The photoacoustic spectrum was obtained using a double-beam spectrometer having carbon black in the reference channel. The dissociation limit is shown by the arrow. (Reproduced by permission from Harshbarger and Robin, 1973a.)

Energetically then one is investing 72 kcal/mole to produce cold O and NO fragments, which can then react further to release 47 kcal/mole. Thus even if all the photon energy at 4000 Å is used to photodissociate the NO_2, there is a release of $\frac{47}{72}$ of this energy in the subsequent reaction. One therefore expects that the photoacoustic signal be reduced by only $\frac{25}{72}$, or about 30%, at 4000 Å, and this expectation is in agreement with the experimental data.

7.4 INTERSYSTEM CROSSING

The biacetyl molecule, a diketone, displays two $n \rightarrow \pi^*$ transitions in the 200–500 nm region, the $S_0 \rightarrow S_1$ transition at 420 nm and the $S_0 \rightarrow S_2$ transition at 290 nm. In Figure 7.3 the photoacoustic spectrum of biacetyl vapor is shown (Kaya et al., 1974, 1975). In this experiment the phase angle was chosen so as to maximize the amplitude of the S_2 band, and this then represents the in-phase spectrum. The in-phase spectrum exhibits

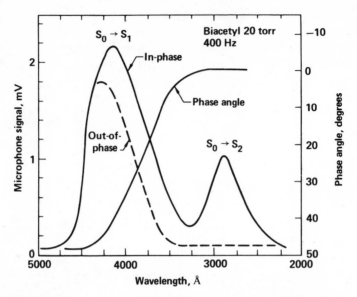

Figure 7.3 The in-phase and out-of-phase photoacoustic spectra of biacetyl, taken with a phase angle chosen to extremize the S_2 signal. The phase angle at which the signal is nulled at each wavelength is also shown. (Reproduced by permission from Kaya et al., 1975.)

both the S_1 and S_2 bands. The quadrature component, that is, the spectrum that is at 90° out-of-phase to the in-phase spectrum, essentially nulls the S_2 band and any other spectral region that is relaxing at the same rate as the S_2 level. The out-of-phase spectrum, shown in Figure 7.3, is null across the entire S_2 band profile, as well as through the short-wavelength end of the S_1 band. However, the long-wavelength end of S_1 is clearly relaxing at a different rate. This is seen more clearly by plotting the phase angle for a null signal versus wavelength. The phase is constant through S_2 but changes continuously through S_1.

From conventional studies, it is known that excitation into S_1 is followed by a very rapid intersystem crossing to a triplet manifold of a neighboring biacetyl molecule, and that this crossing occurs with very high quantum yield. In the region of the $S_1(0,0)$ only T_1 is energetically accessible. The T_1 state has ^3A_u symmetry, and as such, has no spin–orbit matrix element that can assist in relaxing it to the A_g ground state. Hence the radiationless relaxation from T_1 is slow and permits the radiative process, though slow itself ($\tau = 1.53$ msec), to achieve a significant quantum yield. The slow nonradiative relaxation of the T_1 level relative to the fast relaxation of the S_2 level is reflected in the increase in phase angle. As determined from (7.5), this phase angle corresponds to a nonradiative lifetime $\tau_{nr} = 0.46$

msec, which is not very close to 1.53 msec. This discrepancy arises because (7.6) should really be used rather than (7.5) since we are not dealing with a two-state system. The lifetime of 1.53 msec corresponds to $H/L = 0.26$, which agrees with Hunter's value of 0.23 (Hunter and Stock, 1974). For photon energies greater than 440 nm, the second triplet state of biacetyl becomes involved in the relaxation. According to molecular-orbital calculations (Drent el al., 1973), its symmetry is ^3B_g, which is spin–orbit coupled to the ground state and so may relax nonradiatively to the ground state much faster than T_1. Thus as the more energetic parts of the S_1 band are excited, the intersystem crossing to T_2 becomes more favored, with subsequent rapid relaxation to S_0, bypassing T_1 altogether. Finally, in S_2 the relaxation is primarily through photochemical processes, rather than through phosphorescence or radiationless transitions. This fact then accounts for the much smaller amplitude of the S_2 band in the photo-acoustic spectrum.

Intersystem crossings can occur between molecules of different species, as well as between molecules of the same species, as is demonstrated in

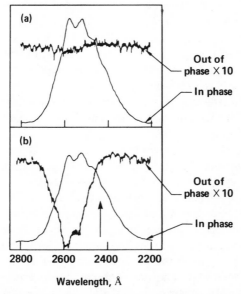

Figure 7.4 Photoacoustic spectrum of (a) benzene vapor (68 torr, 400 Hz) in the absence of biacetyl and (b) benzene vapor (70 torr, 400 Hz) in the presence of 0.3 torr of biacetyl. The arrow in the lower panel indicates the onset of channel 3 relaxation as determined from line-broadening and luminescence studies. (Reproduced by permission from Kaya et al., 1974.)

Figure 7.4, where the photoacoustic spectrum of benzene vapor is shown (Kaya et al., 1974). The lack of an out-of-phase signal for pure benzene vapor indicates that all parts of the $^1A_g \rightarrow {}^1B_{2u}$ absorption band are relaxing either at the same rate or at rates faster than 10^6 sec^{-1}. However, upon the addition of a trace of biacetyl, a slow-heat signal develops, this being the result of an $S \rightsquigarrow T$ intersystem crossing between benzene molecules, followed by energy transfer to the long-lived T_1 state of biacetyl. From a study of the out-of-phase signal it appears that the energy transfer between T_1(benzene) and T_1 (biacetyl) does not extend through the entire $^1B_{2u}$ absorption band, but only across the lower 3000 cm^{-1} of the band. Thus the T_1 state of benzene is accessible only from a limited span of S_1, and at more energetic parts of the S_1 band, another relaxation process appears that is very fast and avoids T_1. This fast relaxation is called "channel 3" and is thought to involve interconversion to an isomeric form of benzene.

7.5 IMPURITY QUENCHING

In the visible spectrum of iodine vapor, there are three overlapping excited states: the $^3\Pi_{1u}$ state, which is only very slightly bound and dissociates into ground state atoms ($^2P_{3/2}$), the unbound $^1\Pi_{1u}$ state, which also dissociates to $^2P_{3/2}$ atoms, and the $^3\Pi_{0u^+}$ excited state, which is strongly bound and dissociates at $\sim$500 nm into one ground state iodine atom and one excited iodine atom ($^2P_{1/2}$). The total absorption spectrum is shown in Figure 7.5a, with the fine-detailed vibronic structure being part of the $^3\Pi_{0u^+}$ band. The photoacoustic spectrum in Figure 7.5b was taken with far less resolution and so reflects only the envelope of the absorption spectrum (Harshbarger and Robin, 1973b).

If small amounts of an impurity gas such as NO are added, there is no discernible change in the PAS spectrum (Figure 7.5c). However, in the presence of oxygen (Figure 7.5d), the PAS spectrum is altered dramatically, showing that, for wavelengths less than 520 nm, a significant fraction of the absorbed energy is not relaxing in a nonradiative mode.

This falloff in the photoacoustic signal occurs at the dissociation limit of the $^3\Pi_{0u^+}$ state. The $^2P_{1/2}$ excited iodine atom produced by this dissociation is 7603 cm^{-1} above the ground state, whereas the oxygen molecule has its first excited state ($^1\Delta_g$) at 7886 cm^{-1} above its $^3\Sigma_g^-$ ground state. Therefore it is quite conceivable that the energy transfer

$$\mathrm{I}\left(^2P_{1/2}\right) + \mathrm{O}_2\left(^3\Sigma_g^-\right) \rightarrow \mathrm{I}\left(^2P_{3/2}\right) + \mathrm{O}_2\left(^1\Delta_g\right) \tag{7.13}$$

occurs since it requires only 284 cm^{-1} of energy. Now the $^1\Delta_g$ state of

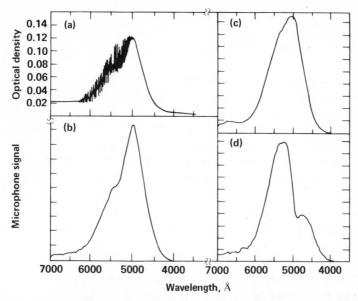

Figure 7.5 (a) The optical absorption spectrum of iodine vapor. (b) The corresponding photoacoustic signal (400 Hz). (c) Photoacoustic signal of I_2 (0.37 torr) and NO (25 torr) gas mixture (400 Hz). (d) Photoacoustic signal of I_2 (0.37 torr) and O_2 (25 torr) gas mixture (400 Hz). (Reproduced by permission from Harshbarger and Robin, 1973b.)

oxygen is quite metastable with a radiative lifetime of over 45 min. Even through collisional relaxation with other O_2 molecules, the lifetime of this state at the experimental pressure of 25 torr is 0.4 sec. Thus this state relaxes too slowly to appear in the PAS experiment, which accounts for the drop in PAS signal below 520 nm. Not all the PAS signal below 520 nm is lost, however, since up to 50% of the dissociated iodine atoms are in the $^2P_{3/2}$ ground state (Oldman et al., 1971).

Such impurity quenching of the photoacoustic signal requires that the impurity have an excited level energetically close to the excited level of the absorbing molecule, and that the lifetime of the impurity level be long with respect to $1/\omega$.

Oxygen is an excellent quencher in that regard, not only for iodine, but also for biacetyl, where it quenches the phosphorescence near the $S_1(0,0)$ band region. The quenching here is thought to occur by a very fast photooxidation process. Thus one would also expect that in the presence of oxygen any of the slow heat observed in the photoacoustic spectrum of the S_1 band would be quenched, and only fast heat would occur. This is exactly what is observed (Kaya et al., 1974), with no out-of-phase signal appearing in the PAS spectrum.

REFERENCES

Cottrell, T. L., Macfarlane, I. M., Read, A. W., and Young, A. H. (1966). *Trans. Faraday Soc.* **62**, 2655.

Delaney, M. E. (1959). *Sci. Prog.* **47**, 459.

Drent, E., van der Werf, R. P., and Kommandeur, J. (1973). *J. Chem. Phys.* **59**, 2061.

Gorelik, G. (1946). *Dokl. Akad. Nauk. SSSR* **54**, 779.

Harshbarger, W. R., and Robin, M. B. (1973a). *Acc. Chem. Res.* **6**, 329.

Harshbarger, W. R., and Robin, M. B. (1973b). *Chem. Phys. Lett.* **21**, 462.

Hunter, T. F., Rumbles, D., and Stock, M. G. (1974). *J. Chem. Soc. Faraday Trans. II* **70**, 1010.

Kaya, K., Harshbarger, W. R., and Robin, M. B. (1974). *J. Chem. Phys.* **60**, 4231.

Kaya, K., Chatelain, C. L., Robin, M. B., and Kuebler, N. A. (1975). *J. Am. Chem. Soc.* **97**, 2153.

Oldman, R. J., Sander, R. K., and Wilson, K. R. (1971). *J. Chem. Phys.* **54**, 4127.

Read, A. W. (1967–1968). *Adv. Mol. Relaxation Processes* **1**, 257.

Slobodskya, P. V. (1948). *Izv. Akad. Nauk SSSR Ser. Fiz.* **12**, 656.

OTHER GAS PAS EXPERIMENTS

8.1 EXCITED-STATE SPECTROSCOPY

At normal laboratory temperatures, the populations of excited molecular states, even if they are low-lying vibrational states, are so low that conventional infrared spectroscopic investigation of these levels presents serious difficulties. With photoacoustic spectroscopy it is now possible to measure infrared absorption coefficients as small as 10^{-10} cm^{-1} when a low-field spin-flip Raman laser is used (Patel, 1971).

Zharov et al. (1977) employed a 1-W CO_2 laser at 9.6 μm to investigate the excited vibrational states of CO_2, BCl_3, and BF_3 molecules. To obtain the energy level for the excited vibrational states, they made use of the fact that the absorption coefficient from a level of energy E is given by

$$\beta(E) = \sigma n(E) = \frac{\sigma n_0 g e^{-E/kT}}{Z_{\text{rot}} + Z_{\text{vib}}} \qquad (8.1)$$

where σ is the transition cross section, g is the level degeneracy, n_0 is the gas density, and Z_{rot} and Z_{vib} are the rotational and vibrational statistical sums. Assuming harmonic oscillator and rigid rotator approximations, one can then derive for a linear polyatomic molecule (Herzberg, 1945),

$$\frac{d\ln(\beta)}{d(1/kT)} = -E + kT + \sum_i \left(\frac{\hbar\omega_i}{e^{\hbar\omega_i/kT-1}} \right) \qquad (8.2)$$

where ω_i is the frequency of the ith vibration.

Using (8.1), Zharov et al. measured the values of the first two vibrational levels in CO_2, BCl_3, and BF_3 by changing the temperature from room temperature to about 250°C. The results for CO_2 are within the limits of error of more conventional experiments, while those for BCl_3 and BF_3 are somewhat more ambiguous.

Patel et al. (1977) used a InSb spin-flip Raman laser to perform spectroscopic studies on the $v=1\rightarrow2$ vibrational–rotational (v–r) transitions of NO. The $v=1$ level is populated by excitation with the $P_{8\text{-}7}$ (11)

line of a fixed-frequency CO laser. The spin-flip Raman laser wavelength is tuned by a magnetic field through a shift in the Zeeman magnetic sublevels in InSb. From this study Patel et al. were able to accurately measure a total of seven $v = 1 \rightarrow 2$ transitions of the $^2\Pi_{1/2}$ state and six $v = 1 \rightarrow 2$ transitions of the $^2\Pi_{3/2}$ states. They were able to assign all the various transitions, obtain the band centers of both the $^2\Pi_{1/2}$ and $^2\Pi_{3/2}$ states, and calculate the anharmonicity for NO for the $^2\Pi_{3/2}$ substate.

Furthermore, by performing this study with PAS the cell used was only 1 cm long, whereas previous excited state spectroscopy on NO by conventional transmission methods required cells 60 cm long (Guerra et al., 1977). With the much smaller cell, Patel et al. were able to insert the cell into the gap of a superconducting solenoid and perform, for the first time, a study of the Zeeman spectra of the excited states of NO.

In a subsequent experiment Patel (1978) was able to perform the same type of measurements on the $v = 1 \rightarrow 2$ excited state transitions of ^{15}NO, in spite of the fact that there is no source available for direct excitation of the $v = 1$ level of ^{15}NO. Patel made use of selective vibrational energy transfer between ^{14}NO excited into its $v = 1$ level by the P_{8-7} (11) CO laser line and ^{15}NO ($v = 1$). Since the $v = 1$ level of ^{14}NO is only 33 cm^{-1} higher than that for ^{15}NO, the energy transfer from ^{14}NO ($v = 1$) to ^{15}NO ($v = 0$) to produce ^{14}NO ($v = 0$) and ^{15}NO($v = 1$) can be quite efficient at room temperature. In addition to obtaining spectroscopic data on the excited states of ^{15}NO, Patel was also able to obtain the vibrational energy transfer rate between the two isotopic species.

8.2 HIGH-RESOLUTION SPECTROSCOPY

The use of lasers in spectroscopy considerably increases the resolution and sensitivity of chemical and isotopic analysis of molecular species, which combined with the high sensitivity for recording weak absorptions with the photoacoustic effect, makes it possible to perform high-resolution optical spectroscopy on gas mixture with species concentrations as low as 10 ppb. In addition to recording absolute abundances of various stable isotopes such as ^{12}C, ^{13}C, ^{16}O, ^{17}O, ^{14}N, and ^{15}N, a laser photoacoustic spectrometer can also determine the ratios of small concentrations of ^{12}C and ^{13}C, and of ^{10}B and ^{11}B, and so on.

Gomenyuk et al. (1976) employed a two-cell differential photoacoustic spectrometer to determine the relative concentrations of ^{10}BCl$_3$ and ^{11}BCl$_3$ molecules admixed in an oxygen environment. The ability to make precise measurements of isotopic ratios makes it possible to measure laser isotope enrichment processes. Gomenyuk et al. have estimated that if the laser

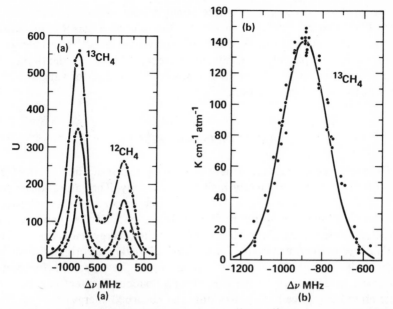

Figure 8.1 (*a*) Photoacoustic spectra of the mixture $^{12}CH_4/^{13}CH_4/air = 1:1.65:660$ taken at pressures from 10 to 20 torr. (*b*) Photoacoustic spectrum (·) of $^{13}CH_4$. Solid curve represents computed Doppler contour. (Reproduced by permission from Zuev et al., 1978.)

radiation can be stabilized, a differential photoacoustic system, similar to theirs, could measure changes in isotopic ratios as small as $10^{-4}\%$.

The high sensitivity of photoacoustic spectroscopy for absorption also can be used to advantage when one is working at low light levels and high resolution for small sample sizes. For example, Zuev et al. (1978) used a gas photoacoustic system to obtain the separate absorption bands of $^{13}CH_4$ in a mixture containing $^{12}CH_4/^{13}CH_4/air = 1:1.65:660$. Figure 8.1 shows their results. The wavelength tuning is achieved by Zeeman shifting the 3.39 μm line of a He–Ne laser. Even though the mean intensity of the Zeeman component was only 1–7 μW, and the sample pressure was only 10–20 torr, the photoacoustic results are quite good.

8.3 PHOTOLYSIS

There is, in general, no suitable conventional spectroscopic method for studying the steady-state intermediate products produced during continuous photolysis. This is generally true of most photolysis experiments since the steady-state concentrations of these intermediates are usually too small for absorption spectroscopy unless some matrix isolation procedure is used. Colles et al. (1976) reported on the use of a gas photoacoustic spectrometer to detect intermediate concentrations of 10^{13} molecules/cm^3 (ppm) during the photolysis of nitromethane. In addition to identifying the intermediate species, Colles et al. also studied the actions of various buffer gases in the photolytic reaction.

8.4 NONLINEAR EFFECTS

In Chapter 3 we discuss the situation when optical saturation occurs for a two-level system at sufficiently high laser intensity. The saturation energy of a quantum transition in a laser amplifier or an absorber is usually measured by determining the nonlinear dependence of the energy leaving the investigated medium with respect to the energy entering it. This method is, however, unsuitable for the case of weakly absorbing transitions because the change in transmitted energy upon the onset of saturation is too small to detect. In such cases, the photoacoustic effect can be used quite effectively since it measures only the absorbed energy.

Ryabov (1975) used this technique for determining the saturation energy of both CO_2 and SF_6 gases that were strongly diluted in argon. The equation for the propagation of a pulse in a two-level resonant medium (Kryukov and Letokhov, 1970) shows that in the homogeneous broadening case when the pulse duration τ_p is much smaller than the relaxation time T_1 of the upper level, the input and output energies are related by

$$\ln\left(\frac{1-e^{-E_0/E_s}}{1-e^{-E_i/E_s}}\right)=\beta_0 l \tag{8.3}$$

where E_0, E_i, and E_s are the output, input, and saturation energies, respectively, β_0 is the linear absorption coefficient, and l is the cell length.

When the absorption is low,

$$E_0 \simeq E_i = E \tag{8.4}$$

and (8.3) reduces to

$$\frac{1}{E}+\frac{1}{2E_s}=\frac{\beta_0 l}{\Delta E} \tag{8.5}$$

where E is the energy density in a pulse and $\Delta E = E_i - E_0 =$ energy absorbed. Thus a plot of the dependence of $1/\Delta E$ on $1/E$, gives a straight line that intersects the ordinate at $1/2E_s$.

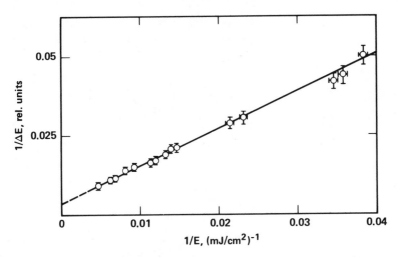

Figure 8.2   Dependence of $1/\Delta E$ on $1/E$ for CO_2 in a mixture of 20 mm Hg CO_2 and 1 atm Ar. (Reproduced by permission from Ryabov, 1975.)

Figure 8.2 shows the experimentally obtained dependence of $1/\Delta E$, that is, $1/q$, where q is the photoacoustic signal normalized in energy units, on $1/E$ for a mixture of 20 mm Hg CO_2 and 1 atm Ar. Under these conditions the absorption coefficient of the signal is 4×10^{-5} cm^{-1}. The experimental dependence is linear and the extrapolation to the ordinate gives a saturation energy E_s for CO_2 of 150 mJ/cm^2, which is in reasonable agreement with the predicted value of 120 mJ/cm^2.

The SF_6 molecule is of considerable interest to laser spectroscopists since it has been shown to undergo collisionless multiphoton dissociation by pulsed CO_2 laser radiation (Ambartzumian et al., 1975; Lyman et al., 1975). Some of these dissociation studies have been performed with photoacoustic systems (Bagratashvili et al., 1975; Black et al., 1977). Conventional energy absorption measurements are difficult to perform at the low gas densities required for collision-free excitation and dissociation. Deutsch (1977) has also used a photoacoustic method to study SF_6 excitation below the dissociation level. He has extended the pulsed photoacoustic technique to low temperature, 145°K, where the hot-band contributions to the absorption of SF_6 are essentially eliminated. In addition to observing two-level saturation effects, Deutsch also studied the effect of collisions during the laser pulse. When the pulse time is comparable to the $v-v$ relaxation time, the relaxation processes occurring during the laser pulse can alter the energy deposition processes, as shown in Figure 8.3, in which the PAS signal from SF_6 at 198°K is plotted as a function of CO_2 laser frequency for pulses of equal energy but different length. The PAS

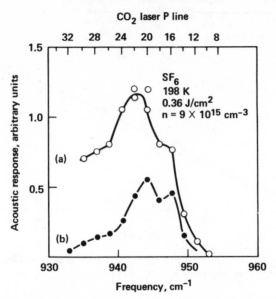

Figure 8.3 Photoacoustic spectra of SF_6 at 198°K for (a) a 6-μsec-long CO_2 laser pulse; (b) a 180-nsec-long pulse. (Reproduced by permission from Deutsch, 1977.)

spectrum for the 6-μsec pulse shows a much more pronounced low-frequency tail, as well as a greater overall absorption. Similar results were also observed by Basov et al. (1976). Deutsch suggests that the change in absorption spectrum is due to the formation of hot bands when the laser-excited ν_3 mode, pumped off-resonance decays into lower-energy modes, producing hot bands, such as $\nu_6 + \nu_3$, that absorb to the red side of ν_3. Such laser-induced hot bands may absorb significantly whenever the laser pulse length is comparable to the v-v relaxation time and the pumped mode can decay into lower-energy modes.

Cox (1978) also applied the photoacoustic method to SF_6. In this study, he was able to investigate linear absorption, saturation, and the onset of multiphoton excitation at intensities below the dissociation limit. His results are shown in Figure 8.4. For the upper curve of Figure 8.4, the gas mixture is 0.1 torr SF_6 and 89 torr Ar. In the lower curve the SF_6 pressure is kept at 0.1 torr, but the argon pressure is lowered to 9 torr. In the upper curve there is linear absorption at intensities below 1 kW/cm², followed by saturation for laser intensities up to 5 MW/cm². But in the lower curve of Figure 8.4, obtained with 10 times less buffer gas, a new intensity dependence is observed above 150 kW/cm². This increased absorption is due to the onset of multiphoton excitation of SF_6.

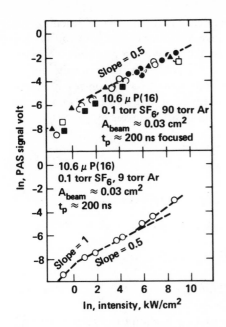

Figure 8.4 The photoacoustic signal for SF_6 is shown as a function of laser intensity for two different buffer gas pressures. Multiphoton excitation occurs for $I_0 > 150$ kW/cm^2 for the lower buffer gas pressure. (Reproduced by permission from Cox, 1978.)

The basic features of the lower curve of Figure 8.4 can be explained by the following simple analysis. The absorption coefficient β that appears in Beer's law,

$$I_T = Ie^{-\beta l} \tag{8.6}$$

can be expressed for a two-level system as

$$\beta = \frac{\beta_0}{1 + I/I_s} \tag{8.7}$$

where β_0 is the linear absorption coefficient, l is the absorption pathlength, I_T is the transmitted intensity, I is the incident intensity, and I_s is the saturation intensity. It has been shown (Kielmann, 1976) that for a collection of two-level absorbers having a distribution of resonant frequencies that is smooth and broad compared to the Lorentzian width for an individual transition, the absorption coefficient can be expressed as

$$\beta = \frac{\beta_0}{(1 + I/I_s)^{1/2}} \tag{8.8}$$

The physical explanation for this dependence is that power broadening will increase the interaction linewidth since the number of previously off-resonant absorbers that now fall within the absorption profile is increased by a factor proportional to $(1+I/I_s)^{1/2}$ and this then results in an increased absorption. Simultaneously, the saturations of on-resonant absorbers decrease the absorption coefficient by a factor of $(1+I/I_s)^{-1}$.

For small values of βl, the absorbed intensity is given by

$$I_{abs} \simeq I\beta l = \frac{I\beta_0 l}{(1+I/I_s)^{1/2}} \qquad (8.9)$$

In the linear absorption region, $I \ll I_s$ and (8.9) reduces to

$$I_{abs} \simeq I\beta_0 l \qquad (8.10)$$

while for $I \gg I_s$

$$I_{abs} \simeq (II_s)^{1/2}\beta_0 l \qquad (8.11)$$

Thus a linear absorption region followed by a saturation region with an $I^{1/2}$ dependence is predicted in agreement with the experimental observations.

As the intensity is increased even further, multiple-photon excitation becomes possible. Then

$$I_{abs} \simeq I^n \left(\frac{I_s}{I}\right)^{1/2} \beta_0 l \qquad (8.12)$$

where $n = 2, 3, 4, \ldots$ indicates the order of the multiple-photon excitation process. Thus a two-photon excitation exhibits an $I^{3/2}$ dependence, and so on.

The upper curve of Figure 8.4 indicates that collisions quench the multiple-photon excitation process. During the relatively short 200-nsec pulse of the laser, only the fast intramolecular v–v relaxation rate can alter the excitation process, since v–t, and intermolecular r–r and v–v energy transfers are quite slow at the experimental pressures. Cox suggests that a fast intramolecular energy transfer out of the pumped mode will create hot bands with red-shifted excitation. Thus laser radiation resonant with the Q or R branch of ν_3 will be further off resonance for excited state absorption as a result of anharmonicity effects than will excitation in the P branch. In this way v–v intramolecular energy transfer may hinder multiple-photon excitation, particularly on the high-frequency side of an absorption band.

This explanation may not be inconsistent with Deutsch's observation of only minor hot-band absorption for 180-nsec pulses, since Deutsch operated at 198°K rather than room temperature.

8.5 MICROWAVE PHOTOACOUSTIC SPECTROSCOPY

Although almost all work in photoacoustic spectroscopy has involved irradiation of the sample with optical photons, it must be kept in mind that a photoacoustic signal can result from the absorption of any electromagnetic energy or, for that matter, of any energy by the sample. Diebold and McFadden (1976) reported on a photoacoustic experiment in which they observed absorption and relaxation of microwave energy between Zeeman magnetic sublevels of molecular oxygen with a photoacoustic cell in a conventional electron paramagnetic resonance spectrometer.

Although the factors that determine the rate of absorption of energy from the radiation field strongly favor optical transitions over microwave transitions, a sufficiently strong microwave absorption ($\sim$mW) is possible because of the high powers and narrow linewidths available from Klystron sources.

In the above experiment, an acoustically resonant photoacoustic cell was centered within the EPR cavity, and the EPR spectrometer was operated in a conventional manner. The cell was filled with 17 torr of oxygen and the spectrometer was swept across the $O_2 E$-line. The microwave absorption was modulated at 1000 Hz by a small a.c. magnetic field superimposed on the d.c. field. Figure 8.5 shows both the conventional diode-detected EPR

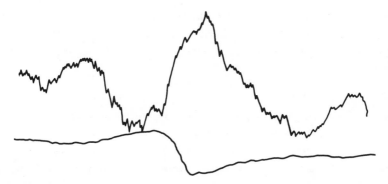

Figure 8.5 (Upper trace) 200-G EPR scan of the O_2 E line centered at 5990 g in 17 torr of oxygen. (Lower trace) photoacoustic signal recorded under identical conditions. (Reproduced by permission from Diebold and McFadden, 1976.)

signal and the PAS signal. The relatively lower signal/noise seen in the PAS scan is partly explained by the weak acoustic resonance of the cell used ($Q\sim 15$), the low density of the O_2, the high modulation frequency, and the acoustic noise emanating from the field modulation coils as a result of vibrations induced by the a.c. magnetic fields. Nevertheless, this experiment clearly demonstrates the possibility of performing valuable photoacoustic experiments in the microwave region of the electromagnetic spectrum.

REFERENCES

Ambartzumian, R. V., Gorokhov, Yu. A., Letokhov, V. S., and Makarov, G. N. (1975). *JETP Lett.* **21**, 171.

Bagratashvili, V. N., Knyazev, I. N., Letokhov, V. S., and Lobko, V. V. (1975). *Opt. Commun.* **14**, 426.

Basov, N. G., Galochkin, V. T., Oraevskii, A. N., and Starodubtsev, N. F. (1976). *JETP Lett.* **23**, 521.

Black, J. G., Yablonowitch, E., and Bloembergen, N. (1977). *Phys. Rev. Lett.* **38**, 1131.

Colles, M. J., Angus, A. M., and Marinero, E. E. (1976). *Nature* **262**, 681.

Cox, D. M. (1978). *Opt. Commun.* **24**, 336.

Deutsch, T. F. (1977). *Opt. Lett.* **1**, 25.

Diebold, G., and McFadden, D. L. (1976). *Appl. Phys. Lett.* **29**, 447.

Gomenyuk, A. S., Zharov, V. P., Letokhov, and Ryabov, E. A. (1976). *Sov. J. Quant. Elect.* **6**, 195.

Guerra, M. A., Sanchez, A., and Javan, A. (1977). *Phys. Rev. Lett.* **38**, 482.

Herzberg, G. (1945). *Infrared and Raman Spectra of Polyatomic Molecules*, Van Nostrand, New York.

Kielmann, F. (1976). *IEEE J. Quant. Elect.* **QE12**, 592.

Kryukov, P. G., and Letokhov, V. S. (1970). *Sov. Phys.-Usp.* **12**, 641.

Lyman, J. L., Jensen, R. J., Rink, J., Robinson, C. P., and Rockwood, S. D., (1975). *Appl. Phys. Lett.* **27**, 87.

Patel, C. K. N. (1971). *Appl. Phys. Lett.* **19**, 400.

Patel, C. K. N. (1978). *Phys. Rev. Lett.* **40**, 535.

Patel, C. K. N., Kerl, R. J., and Burkhardt, E. G. (1977). *Phys. Rev. Lett.* **38**, 1204.

Ryabov, E. A. (1975). *Sov. J. Quant. Elect.* **5**, 81.

Zharov, V. P., Letokhov, V. S., and Ryabov, E. A. (1977). *Appl. Phys.* **12**, 15.

Zuev, V. E., Antipov, A. B., and Sapozhnikova, V. A. (1978). *J. Chem. Phys.* **68**, 1315.

GENERAL THEORY OF THE PHOTOACOUSTIC EFFECT IN CONDENSED MEDIA: THE GAS-MICROPHONE SIGNAL

9.1 INTRODUCTION

In Chapter 2 we discuss some of the theories proposed in the nineteenth century to account for the photoacoustic effect in nongaseous media. These theories range from Bell's concept of air expulsion from pores in the solid surface during heating (Bell, 1881) to Rayleigh's contention that the signal derived primarily from a thermally induced mechanical vibration of the solid (Rayleigh, 1881), and finally to the suggestions of Mercadier (1881) and Preece (1881), that the acoustic signal arises from the periodic heating of the gas in contact with the solid sample.

Experiments performed during the last few years indicate that the primary source of a photoacoustic signal from a condensed sample, as measured by the gas-microphone method, arises from the periodic heat flow from the sample to the surrounding gas with the subsequent change in the gas pressure within the cell. Nevertheless, although Mercadier and Preece were closest to the truth, the mechanisms invoked by Rayleigh and Bell also contribute to the signal, as we see later.

The modern theory of photoacoustics in nongaseous samples is still not complete, although considerable progress has been made during recent years. The first attempt at a modern, quantitative theory was made in 1973 by Parker who, while performing photoacoustic experiments with gases, noticed a small but measurable photoacoustic signal apparently emanating from his cell windows. He then derived the theoretical PAS signal that would be observed from weak absorptions in essentially transparent windows. Although his theory applied to this special case only, many of the salient features for the more general PAS theories can be found in his treatment. As an interesting sidenote, Parker found that his experimental results could only be explained by assuming an anomalously large optical absorption at the surface of his windows, a finding that has now been accepted as generally valid for most polished surfaces of "transparent" windows.

Some years later a more general theory for the photoacoustic effect in condensed media was formulated by Rosencwaig and Gersho (1975; 1976). This theory, now commonly referred to as the RG theory, shows that in a gas-microphone measurement of a PAS signal, the signal depends both on the generation of an acoustic pressure disturbance at the sample–gas interface and on the transport of this disturbance through the gas to the microphone. The generation of the surface pressure disturbance depends in turn on the periodic temperature at the sample–gas interface. The RG theory derived exact expressions for this temperature, while it treated the transport of the disturbance in the gas in an approximate heuristic manner, which is, however, valid for most experimental conditions.

In the following years, further improvements were made. Bennett and Forman (1976), Aamodt et al. (1977), and Wetsel and McDonald (1978) refined the theory by treating the transport of the acoustic disturbance in the gas more exactly, with Navier-Stokes equations. Although these refinements did not change the basic results of the RG theory for most experimental conditions, they were able to account for observed deviations from the RG theory at very low frequencies and at frequencies near the cell resonances. There has been a further refinement to the theory by McDonald and Wetsel (1978), who have included contributions to the signal from thermally induced vibrations in the sample.

In this chapter we present the RG theory and consider some of the later improvements to it.

9.2   ROSENCWAIG-GERSHO THEORY

9.2.1   The Thermal Diffusion Equations

We start with the Rosencwaig-Gersho theory, a one-dimensional analysis of the production of a photoacoustic signal in a simple cylindrical cell such as the one depicted in Figure 9.1. The photoacoustic cell has a diameter D and length L. We assume that the length L is small compared to the wavelength of the acoustic signal, and the microphone (not shown) detects the average pressure produced in the cell. The sample is considered to be in the form of a disc having diameter D and thickness l. The sample is mounted so that its back surface is against a poor thermal conductor of thickness l''. The length l' of the gas column in the cell is then given by $l' = L - l - l''$. We further assume that the gas and backing materials are not light absorbing.

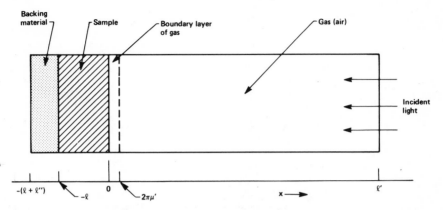

Figure 9.1 Cross-sectional view of a simple cylindrical photoacoustic cell. (Reproduced by permission from Rosencwaig and Gersho, 1976.)

We define the following parameters:

κ: the thermal conductivity (cal/cm-sec-°C)
ρ: the density (g/cm³)
C: the specific heat (cal/g-°C)
$\alpha = \kappa/\rho C$: the thermal diffusivity (cm²/sec)
$a = (\omega/2\alpha)^{1/2}$: the thermal diffusion coefficient (cm⁻¹)
$\mu = 1/a$: the thermal diffusion length (cm)

where ω denotes the chopping frequency of the incident light beam in radians per second. In the following treatment, we denote sample parameters by unprimed symbols, gas parameters by singly primed symbols, and backing material parameters by doubly primed symbols. Table 9.1 lists the above photoacoustic paramenters for a number of common substances.

We assume a sinusoidally chopped monochromatic light source with wavelength λ incident on the solid with intensity

$$I = \tfrac{1}{2}I_0(1 + \cos \omega t) \tag{9.1}$$

where I_0 is the incident monochromatic light flux (W/cm²). We let β denote the optical absorption coefficient of the solid sample (in reciprocal centimeters) for the wavelength λ. The heat density produced at any point x due to light absorbed at this point in the solid is then given by

$$\tfrac{1}{2}\beta I_0 e^{\beta x}(1 + \cos \omega t) \tag{9.2}$$

TABLE 9.1

Substance	Density ρ (g/cm^3)	Specific heat C (cal/g-°C)	Thermal Conductivity κ(cal/cm-sec-°C)	Thermal Diffusivity $\alpha=\kappa/\rho C$ (cm^2/sec)	Thermal diffusion Length at 100 Hz $\mu=(2\alpha/\omega)^{1/2}$(cm)
Aluminum	2.7	0.216	4.8×10^{-1}	0.82	5.1×10^{-2}
Stainless steel	7.5	0.12	3.3×10^{-2}	3.7×10^{-2}	1.1×10^{-2}
Brass	8.5	0.089	2.6×10^{-1}	0.34	3.3×10^{-2}
KCl crystal	2.0	0.21	2.2×10^{-2}	5.2×10^{-2}	1.3×10^{-2}
Crown glass	2.6	0.16	2.5×10^{-3}	6.0×10^{-3}	4.4×10^{-3}
Quartz	2.66	0.188	2.2×10^{-3}	4.4×10^{-3}	3.7×10^{-3}
Rubber	1.12	0.35	3.7×10^{-4}	9.4×10^{-4}	1.7×10^{-3}
Polyethylene	0.92	0.55	5×10^{-4}	9.9×10^{-4}	1.8×10^{-3}
Water	1.00	1.00	1.4×10^{-3}	1.4×10^{-3}	2.1×10^{-3}
Ethyl alcohol	0.79	0.60	4.2×10^{-4}	8.9×10^{-4}	1.7×10^{-3}
Chloroform	1.53	0.23	2.9×10^{-4}	8.4×10^{-4}	1.6×10^{-3}
Air	1.29×10^{-3}	0.24	5.7×10^{-5}	0.19	2.5×10^{-2}
Helium	1.80×10^{-4}	1.25	3.4×10^{-4}	1.52	7.0×10^{-2}

Source: Rosencwaig, 1978.

where x takes on negative values since the solid extends from $x=0$ to $x=-l$, with the light incident at $x=0$. Note also from Figure 9.1 that the air column extends from $x=0$ to $x=l'$ and the backing from $x=-l$ to $x=-(l+l'')$.

The thermal diffusion equation in the solid taking into account the distributed heat source can be written as

$$\frac{\partial^2\theta}{\partial x^2} = \frac{1}{\alpha}\frac{\partial\theta}{\partial t} - Ae^{\beta x}(1+e^{i\omega t}) \qquad \text{for } -l \leqslant x \leqslant 0 \qquad (9.3)$$

with

$$A = \frac{\beta I_0 \eta}{2\kappa} \qquad (9.4)$$

where θ is the temperature and η is the efficiency at which the absorbed light at wavelength λ is converted to heat by the nonradiative deexcitation processes. We assume $\eta=1$, a reasonable assumption for most solids at room temperature. For the backing and the gas, the heat diffusion equations are respectively given by

$$\frac{\partial^2\theta}{\partial x^2} = \frac{1}{\alpha''}\frac{\partial\theta}{\partial t} \qquad -l''-l \leqslant x \leqslant -l \qquad (9.5)$$

$$\frac{\partial^2\theta}{\partial x^2} = \frac{1}{\alpha'}\frac{\partial\theta}{\partial t} \qquad 0 \leqslant x \leqslant l' \qquad (9.6)$$

The real part of the complex valued solution $\theta(x,t)$ of (9.3)–(9.6) is the solution of physical interest and represents the temperature in the cell relative to ambient temperature as a function of position and time. Thus the actual temperature field in the cell is given by

$$T(x,t) = \operatorname{Re}\theta(x,t) + \varphi_0 \qquad (9.7)$$

where Re denotes the "real part of" and φ_0 is the ambient (room) temperature.

To completely specify the solution of (9.3), (9.5), and (9.6), the appropriate boundary conditions are obtained from the requirement of temperature and heat flux continuity at the boundaries $x=0$ and $x=-l$, and from the constraint that the temperature at the cell walls $x=+l'$ and $x=-l-l''$ is at ambient. The latter constraint is a reasonable assumption for metallic cell walls, but in any case it does not affect the ultimate solution for the acoustic pressure. Finally, we make the assumption that the dimensions of the cell are small enough to ignore convective heat flow in the gas at steady-state conditions.

9.2.2   Temperature Distribution in the Cell

The general solution for $\theta(x, t)$ in the cell neglecting transients can be written as

$$
\theta(x,t)=\begin{cases}
\dfrac{1}{l}(x+l+l'')W_0 + We^{\sigma''(x+l)}e^{i\omega t} & -l-l''\leqslant x\leqslant -l \\[2mm]
b_1+b_2x+b_3e^{\beta x}+(Ue^{\sigma x}+Ve^{-\sigma x}-Ee^{\beta x})e^{i\omega t} & -l\leqslant x\leqslant 0 \\[2mm]
\left(1-\dfrac{x}{l'}\right)F+\theta_0e^{-\sigma'x}e^{i\omega t} & 0\leqslant x\leqslant l'
\end{cases}
$$

$$(9.8)$$

where W, U, V, E, and θ_0 are complex valued constants, b_1, b_2, b_3, W_0, and F are real valued constants, and $\sigma=(1+i)a$ with $a=(\omega/2\alpha)^{1/2}$. In particular it should be noted that θ_0 and W represent the complex amplitudes of the periodic temperatures at the sample–gas boundary ($x=0$) and the sample–backing boundary ($x=-l$), respectively. The d.c solution in the backing and gas already make use of the assumption that the temperature (relative to ambient) is zero at the ends of the cell. The quantities W_0 and F denote the d.c. component of the temperature (relative to ambient) at the sample surfaces $x=-l$ and $x=0$, respectively. The quantities E and b_3, determined by the forcing function in (9.3), are given by

$$
b_3=\frac{-A}{\beta^2} \tag{9.9}
$$

$$
E=\frac{A}{(\beta^2-\sigma^2)}=\frac{\beta I_0}{2\kappa(\beta^2-\sigma^2)} \tag{9.10}
$$

In the general solution (9.8) we omit the growing exponential component of the solutions to the gas and backing material because for all frequencies ω of interest the thermal diffusion length is small compared to the length of the material in both the gas and the backing. That is, $\mu''<l''$ and $\mu'<l'$ ($\mu'\sim0.02$ cm for air when $\omega=630$ rad/sec), and hence the sinusoidal components of these solutions are sufficiently damped so that they are effectively zero at the cell walls. Therefore, to satisfy the temperature constraint at the cell walls, the growing exponential components of the solutions would have coefficients that are essentially zero.

The temperature and flux continuity conditions at the sample surfaces are explicitly given by

$$\theta'(0, t) = \theta(0, t) \tag{9.11a}$$

$$\theta''(-l, t) = \theta(-l, t) \tag{9.11b}$$

$$\kappa' \frac{\partial \theta'}{\partial x}(0, t) = \kappa \frac{\partial \theta}{\partial x}(0, t) \tag{9.11c}$$

$$\kappa'' \frac{\partial \theta''}{\partial x}(-l, t) = \kappa \frac{\partial \theta}{\partial x}(-l, t) \tag{9.11d}$$

These constraints apply separately to the d.c. component and the sinusoidal component of the solution. From (9.11), we obtain for the d.c. components of the solution

$$F_0 = b_1 + b_3 \tag{9.12a}$$

$$W_0 = b_1 - b_2 l + b_3 e^{-\beta l} \tag{9.12b}$$

$$\frac{-\kappa'}{l'} F_0 = \kappa b_2 + \kappa \beta b_3 \tag{9.12c}$$

$$\frac{\kappa''}{l''} W_0 = \kappa b_2 + \kappa \beta b_3 e^{-\beta l} \tag{9.12d}$$

Equations (9.12) determine the coefficients b_1, b_2, b_3, W_0, and F_0 for the time-independent (d.c.) component of the solution. Applying (9.11) to the sinusoidal component of the solution yields:

$$\theta_0 = U + V - E \tag{9.13a}$$

$$W = e^{-\sigma l} U + e^{\sigma l} V - e^{-\beta l} E \tag{9.13b}$$

$$-\kappa' \sigma' \theta_0 = \kappa \sigma U - \kappa \sigma V - \kappa \beta E \tag{9.13c}$$

$$\kappa'' \sigma'' W = \kappa \sigma e^{-\sigma l} U - \kappa \sigma e^{\sigma l} V - \kappa \beta e^{-\beta l} E \tag{9.13d}$$

These quantities together with the expression for E in (9.10) determine the coefficients U, V, W, and θ_0. Hence the solutions to (9.12) and (9.13) allow us to evaluate the temperature distribution (9.8) in the cell in terms of the optical, thermal, and geometric parameters of the system. The explicit solution of θ_0, the complex amplitude of the periodic temperature at the solid–gas boundary $(x=0)$ is given by

$$\theta_0 = \frac{\beta I_0}{2\kappa(\beta^2 - \sigma^2)} \left[\frac{(r-1)(b+1)e^{\sigma l} - (r+1)(b-1)e^{-\sigma l} + 2(b-r)e^{-\beta l}}{(g+1)(b+1)e^{\sigma l} - (g-1)(b-1)e^{-\sigma l}} \right]$$

(9.14)

where

$$b = \frac{\kappa'' a''}{\kappa a} \tag{9.15}$$

$$g = \frac{\kappa' a'}{\kappa a} \tag{9.16}$$

$$r = (1-i)\frac{\beta}{2a} \tag{9.17}$$

and, as is stated earlier, $\sigma = (1+i)a$. Thus (9.14) can be evaluated for specific parameter values, yielding a complex number whose real and imaginary parts, θ_1 and θ_2, respectively determine the in-phase and quadrature components of the periodic temperature variation at the surface $x=0$ for the sample. Specifically, the actual temperature at $x=0$ is given by

$$T(0, t) = \varphi_0 + F_0 + \theta_1 \cos \omega t - \theta_2 \sin \omega t \tag{9.18}$$

where φ_0 is the ambient temperature at the cell walls and F_0 is the increase in temperature due to the steady-state component of the absorbed heat.

9.2.3   Production of the Acoustic Signal

As is stated at the beginning of this chapter, it is our contention that the main source of the acoustic signal arises from the periodic heat flow from the solid to the surrounding gas. The periodic diffusion process produces a periodic temperature variation in the gas as given by the sinusoidal (a.c.) component of the solution (9.8),

$$\theta_{\text{a.c.}}(x, t) = \theta_0 e^{-\sigma' x} e^{i\omega t} \tag{9.19}$$

Taking the real part of (9.19), we see that the actual physical temperature variation in the gas is

$$T_{\text{a.c.}}(x, t) = e^{-a' x} \left[\theta_1 \cos(\omega t - a' x) - \theta_2 \sin(\omega t - a' x) \right] \tag{9.20}$$

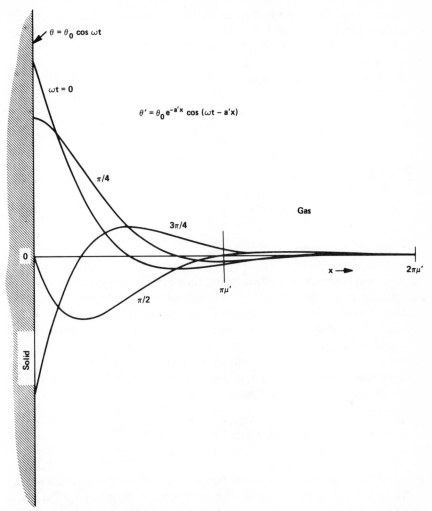

Figure 9.2 Spatial variation of the time-dependent temperature at the gas–sample interface. (Reproduced by permission from Rosencwaig and Gersho, 1976.)

where θ_1 and θ_2 are the real and imaginary parts of θ_0, as given by (9.14). As can be seen in Figure 9.2, the time-dependent component of the temperature in the gas attenuates rapidly to zero with increasing distance from the surface of the solid. At a distance of only $2\pi/a' = 2\pi\mu'$, where μ' is the thermal diffusion length in the gas, the periodic temperature variation in the gas is effectively fully damped out. Thus we can define a boundary layer, as shown in Figure 9.1, whose thickness is $2\pi\mu'(\sim 0.1$ cm at $\omega/2\pi = 100$ Hz) and maintain to a good approximation that only this thickness of gas is capable of responding thermally to the periodic temperature at the surface of the sample.

The spatially averaged temperature of the gas within this boundary layer as a function of time can be determined by evaluating

$$\bar{\theta}(t) = \frac{1}{2\pi\mu'} \int_0^{2\pi\mu'} \theta_{a.c.}(x, t) \, dx \qquad (9.21)$$

From (9.19)

$$\bar{\theta}(t) \simeq \frac{1}{2\sqrt{2}\,\pi} \theta_0 e^{i(\omega t - \pi/4)} \qquad (9.22)$$

using the approximation $e^{-2\pi} \ll 1$.

Because of the periodic heating of the boundary layer, this layer of gas expands and contracts periodically and thus can be thought of as acting as an acoustic piston on the rest of the gas column, producing an acoustic pressure signal that travels through the entire gas column. A similar argument has been used successfully to account for the acoustic signal produced when a conductor in the form of a thin flat sheet is periodically heated by an a.c. electrical current (Arnold and Crandall, 1917).

The displacement of this gas piston due to the periodic heating can be simply estimated by using the ideal gas law,

$$\delta x(t) = 2\pi\mu' \frac{\bar{\theta}(t)}{T_0} = \frac{\theta_0 \mu'}{\sqrt{2}\,T_0} e^{i(\omega t - \pi/4)} \qquad (9.23)$$

where we have set the average d.c. temperature of this gas boundary layer equal to the d.c. temperature at the solid surface, $T_0 = \varphi_0 + F_0$. Equation (9.23) is a reasonable approximation to the actual displacement of the layer since $2\pi\mu'$ is only ~ 0.1 cm for $\omega/2\pi = 100$ Hz and even smaller at higher frequencies.

If we assume that the rest of the gas responds to the action of this piston adiabatically, then the acoustic pressure in the cell due to the displacement of this gas piston is derived from the adiabatic gas law

$$PV^\gamma = \text{constant} \qquad (9.24)$$

where P is the pressure, V is the gas volume in the cell, and γ is the ratio of the specific heats. Thus the incremental pressure is

$$\delta P(t) = \frac{\gamma P_0}{V_0} \delta V = \frac{\gamma P_0}{l'} \delta x(t) \tag{9.25}$$

where P_0 and V_0 are the ambient pressure and volume, respectively, and $-\delta V$ is the incremental volume. Then from (9.23),

$$\delta P(t) = Q e^{i(\omega t - \pi/4)} \tag{9.26}$$

where

$$Q = \frac{\gamma P_0 \theta_0}{\sqrt{2} \, l'a'T_0} \tag{9.27}$$

Thus the actual physical pressure variation, $\Delta p(t)$, is given by the real part of $\delta \mathcal{P}(t)$ as

$$\Delta p(t) = Q_1 \cos\left(\omega t - \frac{\pi}{4}\right) - Q_2 \sin\left(\omega t - \frac{\pi}{4}\right) \tag{9.28}$$

or

$$\Delta p(t) = q \cos\left(\omega t - \psi - \frac{\pi}{4}\right) \tag{9.29}$$

where Q_1 and Q_2 are the real and imaginary parts of Q, and q and $-\psi$ are the magnitude and phase of Q, that is,

$$Q = Q_1 + iQ_2 = q e^{-i\psi} \tag{9.30}$$

Thus Q specifies the complex envelope of the sinusoidal pressure variation. Combining (9.14) and (9.27) we get the explicit formula

$$Q = \frac{\beta I_0 \gamma P_0}{2\sqrt{2} \, T_0 \kappa l' a' (\beta^2 - \sigma^2)}$$

$$\times \left[\frac{(r-1)(b+1)e^{\sigma l} - (r+1)(b-1)e^{-\sigma l} + 2(b-r)e^{-\beta l}}{(g+1)(b+1)e^{\sigma l} - (g-1)(b-1)e^{-\sigma l}} \right] \tag{9.31}$$

where $b = \kappa''a''/\kappa a$, $g = \kappa'a'/\kappa a$, $r = (1-i)\beta/2a$, and $\sigma = (1+i)a$ as is defined earlier. At ordinary temperatures $T_0 \simeq \varphi_0$ so that the d.c. components of the temperature distribution need not be evaluated. Thus (9.31) may be evaluated for the magnitude and phase of the acoustic pressure wave produced in the cell by the photoacoustic effect.

9.3 SPECIAL CASES

The full expression for $\Delta p(t)$ is somewhat difficult to interpret because of the complicated expression of Q as given by (9.31). However, physical insight may be gained by examining special cases where the expression for Q becomes relatively simple. We group these cases according to the optical opaqueness of the solids as determined by the relation of the optical absorption length $l_\beta = 1/\beta$ to the thickness l of the solid. For each category of optical opaqueness, we then consider three cases according to the relative magnitude of the thermal diffusion length μ, as compared to the physical length l and the optical absorption length l_β. For all the cases evaluated below, we make use of the reasonable assumption that $g < b$ and that $b \sim 1$, this is, that $\kappa' a' < \kappa'' a''$ and $\kappa'' a'' \simeq \kappa'' a$.

The six cases are shown in Figure 9.3. It is convenient to define

$$Y = \frac{\gamma P_0 I_0}{2\sqrt{2}\, T_0 l'} \tag{9.32}$$

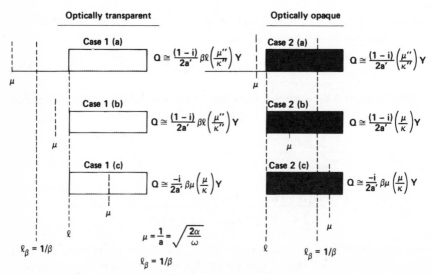

Figure 9.3 Schematic representation of the special cases for the photoacoustic theory of solids. (Reproduced by permission from Rosencwaig and Gersho, 1976.)

which always appears in the expression for Q as a constant factor. We also define the optical path length as

$$l_\beta = \frac{1}{\beta} \qquad (9.33)$$

9.3.1 Optically Transparent Solids ($l_\beta > l$)

In these cases, the light is absorbed throughout the length of the sample, and some light is transmitted through the sample.

Case 1a: Thermally Thin Solids ($\mu \gg 1$; $\mu > l_\beta$)

Here we set $e^{-\beta l} \simeq 1 - \beta l$, $e^{\pm\sigma l} \simeq 1$, and $|r| > 1$ in (9.31). We then obtain

$$Q = \frac{Y}{2a'a''\kappa''}(\beta - 2ab - i\beta) \simeq \frac{(1-i)\beta l}{2a'}\left(\frac{\mu''}{\kappa''}\right)Y \qquad (9.34)$$

The acoustic signal is thus proportional to βl, and since μ''/a' is proportional to $1/\omega$, the acoustic signal has an ω^{-1} dependence. For this thermally thin case of $\mu \gg l$, the thermal properties of the backing material come into play in the expression for Q.

Case 1b: Thermally Thin Solids ($\mu > 1$; $\mu < l_\beta$)

Here we set $e^{-\beta l} \simeq 1 - \beta l$, $e^{\pm\sigma l} \simeq (1 \pm \sigma l)$, and $|r| < 1$ in (9.31). We then obtain

$$Q = \frac{\beta l Y}{4\kappa a'a^3 b}\left[(\beta^2 + 2a^2) + i(\beta^2 - 2a^2)\right] \simeq \frac{(1-i)\beta l}{2a'}\left(\frac{\mu''}{\kappa''}\right)Y \quad (9.35)$$

The acoustic signal is again proportional to βl, varies as ω^{-1}, and depends on the thermal properties of the backing material. Equation (9.35) is identical to (9.34).

Case 1c: Thermally Thick Solids ($\mu < 1$; $\mu \ll l_\beta$)

In (9.31) we set $e^{-\beta l} \simeq 1 - \beta l$, $e^{-\sigma l} \simeq 0$, and $|r| \ll 1$. The acoustic signal then becomes

$$Q \simeq -i\frac{\beta\mu}{2a'}\left(\frac{\mu}{\kappa}\right)Y \qquad (9.36)$$

Here the signal is proportional to $\beta\mu$ rather than βl. That is, only the light absorbed within the first thermal diffusion length contributes to the signal, in spite of the fact that light is being absorbed throughout the length l of the solid. Also, since $\mu < l$, the thermal properties of the backing material present in (9.35) are replaced by those of the sample. The frequency dependence of Q in (9.36) varies as $\omega^{-3/2}$.

Cases 1a, 1b, and 1c for the so-called optically transparent sample demonstrate a unique capability of photoacoustic spectroscopy, to wit, the capability of obtaining a depth profile of optical absorption within a sample; that is, by starting at a high chopping frequency we can obtain optical absorption information from only a layer of material near the surface of the solid. For materials with low thermal diffusivity this layer can be as small as 0.1 μm at chopping frequencies of 10,000–100,000 Hz. Then by decreasing the chopping frequency, we increase the thermal diffusion length and obtain optical absorption data further within the material, until at ~5 Hz we can obtain data down to 10–100 μm for materials with low thermal diffusivities and up to 1–10 mm for materials with high thermal diffusivities.

This capability for depth-profile analysis is unique and opens up exciting possibilities in studying layered and amorphous materials and in determining overlay and thin film thicknesses.

9.3.2   Optically Opaque Solids ($l_\beta \ll l$)

In these cases, most of the light is absorbed within a distance that is small compared to l, and essentially no light is transmitted.

Case 2a: Thermally Thin Solids ($\mu \gg 1; \mu \gg l_\beta$)

In (9.31) we set $e^{-\beta l} \simeq 0$, $e^{\pm \sigma l} \simeq 1$, and $|r| \gg 1$. We then obtain

$$Q \simeq \frac{(1-i)}{2a'} \left(\frac{\mu''}{\kappa''} \right) Y \tag{9.37}$$

In this case, we have photoacoustic "opaqueness" as well as optical opaqueness, in the sense that our acoustic signal is independent of β. This would be the case for a very black absorber such as carbon black. The signal is quite strong (it is $1/\beta l$ times as strong as that in case 1a), depends on the thermal properties of the backing material, and varies as ω^{-1}.

Case 2b: Thermally Thick Solids ($\mu < 1; \mu > l_\beta$)

In (9.31) we set $e^{-\beta l} \simeq 0$, $e^{-\sigma l} \simeq 0$, and $|r| > 1$. We obtain

$$Q \simeq \frac{Y}{2a'a\kappa\beta}(\beta - 2a - i\beta) \simeq \frac{(1-i)}{2a'} \left(\frac{\mu}{\kappa} \right) Y \tag{9.38}$$

Equation (9.38) is analogous to (9.37), but the thermal parameters of the backing are now replaced by those of the solid. Again the acoustic signal is independent of β and varies as ω^{-1}.

Case 2c: Thermally Thick Solids $(\mu \ll 1;\ \mu < l_\beta)$

We set $e^{-\beta l} \simeq 0$, $e^{-\sigma l} \simeq 0$, and $|r| < 1$ in (9.31). We obtain

$$Q = \frac{-i\beta Y}{4a'a^3\kappa}(2a - \beta + i\beta) \simeq \frac{-i\beta\mu}{2a'}\left(\frac{\mu}{\kappa}\right)Y \qquad (9.39)$$

This is a very interesting and important case. Optically we are dealing with a very opaque solid $(\beta l \gg 1)$. However, as long as $\beta\mu < 1$ (i.e., $\mu < l_\beta$), this solid is not photoacoustically opaque, since, as in case 1c, only the light absorbed within the first thermal diffusion length, μ, contributes to the acoustic signal. Thus even though this solid is optically opaque, the photoacoustic signal is proportional to β. As in 1c, the signal is also dependent on the thermal properties of the sample and varies as $\omega^{-3/2}$.

9.4 EXPERIMENTAL VERIFICATION

In this section we consider some of the predictions of the Rosencwaig-Gersho theory of the photoacoustic effect in solids, and how these predictions have been borne out by experiment.

One of the most obvious and important predictions of the theory is that the photoacoustic signal is always linearly proportional to the power of the incident photon beam, and that this dependence holds for any sample or cell geometry. This prediction has been found to be fully accurate. In Section 9.3 we show that when the thermal diffusion length in the sample is greater than the optical absorption path length (cases 2a and 2b), the photoacoustic signal is independent of the optical absorption coefficient of the sample. For these cases, therefore, the only term in (9.37) or (9.38) that is dependent on the wavelength of the incident radiation is the light source intensity I_0. Thus it is clear that the photoacoustic spectrum in the case of a photoacoustically opaque sample $(\mu > l_\beta)$ is simply the power spectrum of the light source. This is verified in Figure 9.4, which shows the PAS spectrum of a porous carbon-black sample and also a power spectrum of the same source, as seen by the photoacoustic cell, taken with a silicon diode power meter (Rosencwaig, 1973). In the wavelength region of $\lambda > 400$ nm, where the silicon diode power meter has a flat wavelength response, we see a one-to-one correspondence between the PAS spectrum and the power spectrum. Unlike the silicon diode power meter, however, a PAS

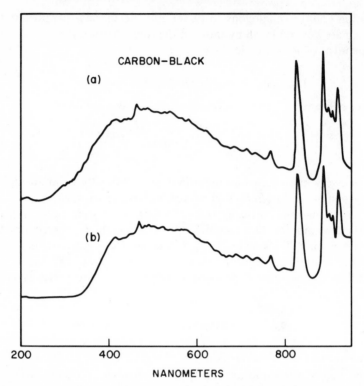

Figure 9.4 (*a*) The photoacoustic spectrum of carbon black. (*b*) The power spectrum of the xenon lamp a silicon diode power taken with meter. (Reproduced by permission from Rosencwaig, 1973.)

cell containing a porous carbon black (e.g., a loose wad of completely burned cotton or a thick layer of carbon-black particles obtained from the incomplete combustion of acetylene) acts as a true light trap with a flat response at all wavelengths. In fact, it is clear from Figure 9.4 that one can readily construct a power meter based on the photoacoustic principle that would have a greater wavelength range than other power meters, while maintaining high sensitivity and a fairly large dynamic range. The same power meter could be used from the X-ray range to the far infrared, requiring only a change of entrance window to permit appropriate transmission of the desired photons into the PAS cell.

A further verification of the validity of the general RG theory has been demonstrated by Adams et al (1979). In this study, the variation of the magnitude of the photoacoustic signal was investigated as a function of

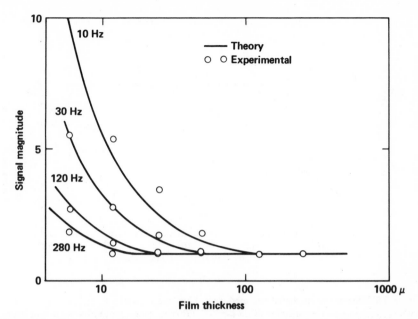

Figure 9.5 The magnitude of the photoacoustic signal versus polymer film thickness for different modulation frequencies. (Reproduced by permission from Adams et al., 1979.)

sample thickness and modulation frequency for thin polymer films. The signal was measured at 300 nm, where the samples were optically thick. The experimental data and the theoretical curves predicted by the RG theory are shown in Figure 9.5. As we see, the agreement is quite good except at very low modulation frequencies, where the thermal thickness of the gas column in front or behind the sample becomes important.

The saturation of the photoacoustic signal that is theoretically expected to occur in an optically opaque material when the thermal diffusion length becomes larger than the optical pathlength has been demonstrated by McClelland and Kniseley (1976) using fixed chopping frequencies and a variable β by working with aqueous solutions of methylene blue dye. Their results are shown in Figure 9.6, where we see that essentially full saturation is reached for a β of ~ 2000 cm^{-1} when the chopping frequency is 50 Hz and for a β of $\sim 30,000$ cm^{-1} when the chopping frequency is 1800 Hz.

One of the most important aspects of the RG theory is that if one has a full knowledge of the thermal and geometrical parameters of the sample, then one can obtain absolute values for the optical absorption coefficient by measuring the dependence of the photoacoustic signal on modulation

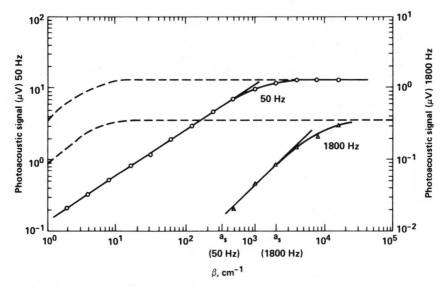

Figure 9.6 Photoacoustic signal for methylene blue dye in water. (_____) Experimental data, (_ _ _ _ _ _) calculated signal with no thermal diffusion. (Reproduced by permission from McClelland and Kniseley, 1976.)

frequency. From (9.31), the photoacoustic signal for a thermally and optically thick sample is given by

$$Q = \frac{\beta\gamma(r-1)P_0 I_0}{2\sqrt{2}\ T_0\kappa l'a'(g+1)(\beta^2-\sigma^2)} \tag{9.40}$$

Wetsel and McDonald (1977a) determined the absorption coefficient for an aqueous phenol red sodium salt solution by fitting the experimental $Q(\beta,\omega)$ to (9.40) as shown in Figure 9.7. The solid curve represents the best fit of (9.40) with the data. The agreement with the calculated value of β from the known concentration of phenol red sodium salt was found to be quite good, thus verifying the RG theory, at least for the frequency range of 100–2000 Hz.

The dynamic range for the absorption coefficient β that can be measured with the photoacoustic effect can be quite large. For example, Monahan and Nolle (1977) have shown that a PAS signal varies significantly with changes in the absorption coefficient β of powdered amorphous As_2S_3 from less than 10 to more than 10^4 cm^{-1} at a fixed chopping frequency of 510 Hz. If the chopping frequency were increased to 5000 Hz,

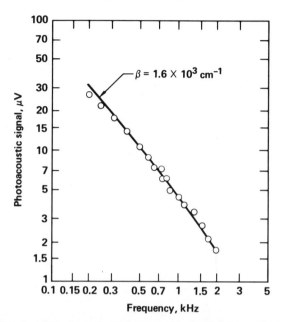

Figure 9.7 Experimentally determined photoacoustic signal (O) as a function of frequency for phenol red sodium salt in distilled water (concentration = 25 g/l). The solid curve represents the best fit of (9.40) to the data: $\beta = 1.6 \times 10^3$ cm^{-1}. (Reproduced by permission from Wetsel and McDonald, 1977.)

a β greater than 10^5 cm^{-1} could be measured. At the low end, we can readily measure absorption coefficients in the range of 10^{-2} cm^{-1} or less if care is taken to eliminate background signals due to window and wall absorptions and to account for diffuse reflection and absorption in powdered samples. Absorption coefficients as low as 10^{-5} cm^{-1} have been measured in nonpowdered solids such as intact alkali fluoride crystals by Hordvik and Schlossberg (1977), who used a piezoelectric variation of the photoacoustic technique. These authors measured the elastic strains that are produced when a crystalline solid is illuminated with high-power periodic laser light by bonding a piezoelectric transducer directly to the material under study. The sensitivity of this technique appears limited only by the amount of radiation scattered directly onto the transducer by impurities and inhomogeneities in the material. Hordvick and Schlossberg estimate that in low light-scattering materials absorption coefficients as low as 10^{-6} cm^{-1} could be measured with laser powers of about 1 W. The total dynamic range for the measurement of β thus appears to be from 10^{-6} to 10^5 cm^{-1}, a range considerably greater than that available with current spectrophotometers.

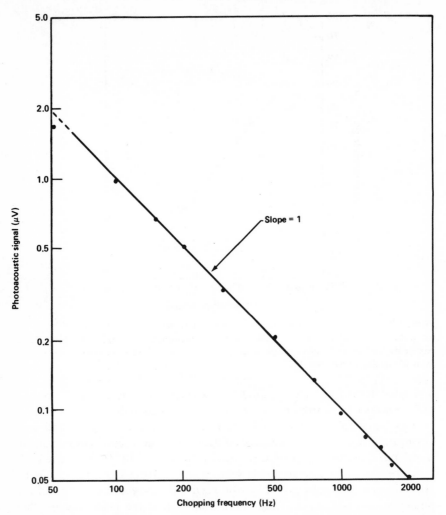

Figure 9.8 A log–log plot of the photoacoustic signal for carbon black versus chopping frequency showing the ω^{-1} dependence. (Reproduced by permission from Rosencwaig, 1978.)

As the Rosencwaig-Gersho theory indicates, the photoacoustic effect is primarily dependent on the relationship between three "length" parameters of the sample: the thickness of the sample l, the optical absorption length $l_\beta = 1/\beta$, and the thermal diffusion length $\mu = (2\alpha/\omega)^{1/2}$.

In the case of a strongly absorbing material, such as fine carbon-black particles 10^{-3} to 10^{-4} cm in diameter, we are dealing with the situation where $l_\beta < l$, since β is of the order of 10^6 cm^{-1}. At the same time, the

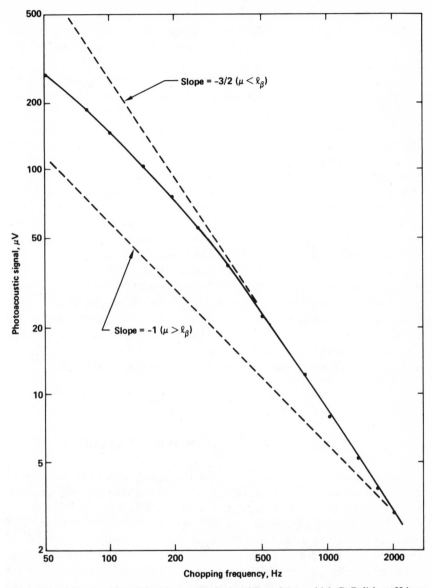

Figure 9.9 A log–log plot of the photoacoustic signal for a 0.1-cm-thick GaP disk at 524 nm versus chopping frequency, showing a frequency dependence that varies from close to ω^{-1}, at low frequencies, to $\omega^{-3/2}$, at high frequencies. (Reproduced by permission from Rosencwaig, 1977.)

113

thermal diffusion length $\mu > l$ for chopping frequencies in the range 50–5000 Hz. The log–log plot of the experimental photoacoustic signal with frequency clearly shows in Figure 9.8 the ω^{-1} behavior predicted by the RG theory (Rosencwaig, 1978).

The theory also predicts that for an opaque material ($l_\beta < l$), the PAS signal will vary as ω^{-1} when $\mu > l_\beta$, and as $\omega^{-3/2}$ when $\mu < l_\beta$. This prediction has been confirmed with an experiment on a disk of GaP 0.1 cm thick (Rosencwaig, 1977). At a wavelength of 524 nm, $\beta \sim 25$ cm^{-1} for GaP at room temperature. Thus by varying the chopping frequency from 50 to 2000 Hz, the theory predicts that we will move from a region where the dependence is primarily ω^{-1} to one where it is $\omega^{-3/2}$. The experimental results shown in Figure 9.9 verify this predicted frequency dependence.

Wetsel and McDonald (1977a) have also shown this change in ω dependence in their photoacoustic study of aqueous solutions of phenol red sodium salt. In fact, knowing the thermal properties of their sample (mainly water), they were able to obtain values for the absorption coefficient accurate to within 10% from an analysis of the frequency dependence of the photoacoustic signal. These authors also noted a deviation from the theory at very low chopping frequencies. This deviation can be accounted for by the fact that at very low frequencies, the assumption made in the RG theory that the gas column in the PAS cell is always much larger than the gaseous thermal diffusion length is, of course, no longer valid. More is said about this point in the theoretical treatment that follows.

9.5   PHOTOACOUSTIC TRANSPORT IN A FLUID

As is shown in the preceding section the RG theory treats the transport of the photoacoustic signal in the gas in an approximate heuristic manner. Aamodt et al. (1977), Bennett and Forman (1976), and McDonald and Wetsel (1978) have all pointed out that this is a weak aspect of the RG theory in that it cannot account for the roll-off in the photoacoustic signal when the gas column is thermally thin ($l' < \mu'$) or for the changes that occur when the gas length approaches an appreciable fraction of the acoustic wavelength. We present below a more exact treatment of the transport process in the gas and, in fact, make it general for any isotropic fluid. This analysis differs somewhat from those in the references above, although the results are the same. We use it because this treatment leads simply and naturally to the analysis of piezoelectrically detected PAS signals presented in Chapter 10.

9.6   THERMOELASTIC THEORY

Following White's (1963) analysis of thermoelastic generation of sound in condensed media, let us assume that a fluid absorbs heat uniformly over its surface at the plane $x=0$. As a result there is a temperature rise at and near the surface and a subsequent strain

$$\varepsilon = \frac{\partial u(x,t)}{\partial x} = \alpha_t \theta(x,t) \qquad (9.41)$$

where $u(x,t)$ is the x component of particle displacement, α_t is the coefficient of linear thermal expansion, $\theta(x,t)$ is the temperature rise above the initial temperature, and t is the time.

In the presence of both heating and actual stress, the usual stress–strain relation

$$p = B\varepsilon \qquad (9.42)$$

becomes

$$p = B\varepsilon - B\alpha_t \theta \qquad (9.43)$$

where B is the bulk modulus. Table 9.2 gives some of the theromelastic parameters for various liquids and solids.

The equation of motion for an elastic wave traveling in the fluid is

$$\rho \frac{\partial^2 u}{\partial t^2} = \frac{\partial p}{\partial x}$$

$$= B\frac{\partial^2 u}{\partial x^2} - B\alpha_t \frac{\partial \theta}{\partial x} \qquad (9.44)$$

Now the sound velocity in the fluid is given by $c_0^2 = B/\rho$. Thus (9.44) becomes

$$\frac{\partial^2 u}{\partial x^2} - \frac{1}{c_0^2}\frac{\partial^2 u}{\partial t^2} = \alpha_t \frac{\partial \theta}{\partial x} \qquad (9.45)$$

If the fluid does not absorb any of the incident light, but is in contact with a light-absorbing sample at the boundary $x=0$, then, as is shown in Section 9.2, the temperature in the fluid is given by

$$\theta(x,t) = \theta_0 e^{-\sigma' x} e^{i\omega t} \qquad (9.46)$$

TABLE 9.2. Thermoelastic Parameters for Various Substances

Substance	Density $g(g/cm^3)$	Coefficient of Linear Thermal Expansion $\alpha_t(\times 10^{-4})$	Bulk Modulus $B(\times 10^{10})(dynes/cm^2)$	Sound Velocity (25° C) $c_0(\times 10^5)(cm/sec)^a$
Liquids				
Acetone	0.790	4.8	0.79	1.20
Carbon disulfide	1.292	4.1	1.08	1.14
Carbon tetrachloride	1.263	4.0	0.94	0.93
Chloroform	1.527	4.2	0.99	1.00
Ethyl alcohol	0.792	3.7	0.90	1.16
Glycerin	1.26	1.7	4.76	1.90
Methyl alcohol	0.791	3.8	0.81	1.12
Water	1.00	0.67	1.89	1.48
Solids				
Aluminum	2.70	0.23	77	6.42
Brass	8.2–8.8	0.18	105	4.70
Copper	8.96	0.17	135	5.01
Glass (crown)	2.5–2.7	0.09	50	5.10
Gold	19.3	0.14	185	3.24
Lead	11.34	0.29	43	1.96
Silver	10.5	0.19	100	3.65
Steel	7.8	0.25	175	5.94

[a]The sound velocities given for solids represent longitudinal velocities.

where θ_0 is the temperature at the sample–fluid interface and is given by (9.31) of the RG theory, and x is positive $0 \leqslant x \leqslant l'$. The general solution for $u(x, t)$ is then

$$u(x, t) = u_0(x)e^{i\omega t} \tag{9.47}$$

Equation (9.45) then can be written as

$$\frac{\partial^2 u}{\partial x^2} + k'^2 u = \alpha'_t \frac{\partial \theta}{\partial x} \tag{9.48}$$

Again, the primed symbols represent parameters of the gas in the cell, where $k' = \omega/c'_0$ is the wave vector for the acoustic wave in the gas. The general solution for $u_0(x)$ is given by

$$u_0(x) = A'e^{-ik'x} + D'e^{ik'x} - \alpha'_t \theta_0 \frac{\sigma'}{(k'^2 + \sigma'^2)} e^{-\sigma'x} \tag{9.49}$$

where A' and D' are constants to be determined from the boundary conditions.

If we are dealing with a gas-microphone, or more generally, a fluid-microphone system, the boundary conditions are that the ends $x = 0$ and $x = l'$ are constrained. Of course, we are assuming that the sample has a greater acoustic impedance than the fluid, a condition certainly true when the fluid is a gas. Thus the boundary conditions are

$$u(0) = 0$$

$$u(l') = 0 \tag{9.50}$$

From (9.49),

$$A' + D' = \alpha'_t \theta_0 \left(\frac{\sigma'}{k'^2 + \sigma'^2} \right) \tag{9.51}$$

$$A'e^{-ik'l'} + D'e^{ik'l'} = \alpha'_t \theta_0 \left(\frac{\sigma'}{k'^2 + \sigma'^2} \right) e^{-\sigma'l'} \tag{9.52}$$

The constants A' and D' are then found to be

$$A' = \alpha'_t \theta_0 \left(\frac{\sigma'}{k'^2 + \sigma'^2} \right) \left(\frac{e^{ik'l'} - e^{-\sigma'l'}}{e^{ik'l'} - e^{-ik'l'}} \right) \tag{9.53}$$

$$D' = \alpha_t \theta_0 \left(\frac{\sigma'}{k'^2 + \sigma'^2} \right) \left(\frac{e^{-ik'l'} - e^{-\sigma'l'}}{e^{ik'l'} - e^{-ik'l'}} \right) \tag{9.54}$$

The strain $\varepsilon'(x)$ and the stress $p'(x)$ are

$$\varepsilon'(x) = \frac{\partial u'}{\partial x} = i\alpha'_t\theta_0\left(\frac{k'\sigma'}{k'^2+\sigma'^2}\right)\left[\frac{(e^{-\sigma'l'}-e^{ik'l'})e^{-ik'x}+(e^{-\sigma'l'}-e^{-ik'l'})e^{ik'x}}{e^{ik'l'}-e^{-ik'l'}}\right]$$

(9.55)

$$p(x) = iB'\alpha'_t\theta_0\left(\frac{k'\sigma'}{k'^2+\sigma'^2}\right)\left[\frac{(e^{-\sigma'l'}-e^{ik'l'})e^{-ik'x}+(e^{-\sigma'l'}-e^{-ik'l'})e^{ik'x}}{e^{ik'l'}-e^{-ik'l'}}\right]$$

$$-B'\alpha'_t\theta_0\left(\frac{k'^2}{k'^2+\sigma'^2}\right)e^{-\sigma'x}$$

(9.56)

We note that when l' is an appreciable fraction of an acoustic wavelength ($k'l'$ not negligibly small), both the amplitude and phase of the PAS pressure p' as measured by a microphone depend on the position x of the microphone. Also, there are cell resonances whenever $k'l'=n\pi$ ($l'=n\lambda'/2$).

When the length l' is much smaller than an acoustic wavelength, $k'l'\ll1$, and we find

$$\varepsilon'(x) \simeq \alpha'_t\theta_0\left[e^{-\sigma'x}-\frac{(1-e^{-\sigma'l'})}{\sigma'l'}\right]$$

(9.57)

$$p'(x) \simeq -B'\alpha'_t\theta_0\left\{\frac{1-e^{-\sigma'l'}}{\sigma'l'}\right\}$$

(9.58)

The stress or pressure in the fluid is thus compressive and independent of position x. When the fluid is thermally thick ($\sigma'l'\gg1$),

$$p' = -\frac{B'\alpha'_t\theta_0}{\sigma'l'}$$

(9.59)

The pressure experienced by the microphone is then equal and opposite to the pressure on the fluid. Now when the fluid is a gas, $B'=\gamma P_0$ and $\alpha'_t=1/T_0$ and the pressure measured by the microphone, that is, the photoacoustic signal, is given by

$$q = -p' = \frac{\gamma P_0}{T_0\sigma'l'}\theta_0 e^{i\omega t}$$

$$= \frac{\gamma P_0}{\sqrt{2}\,T_0 l'a'}\theta_0 e^{i(\omega t-\pi/4)}$$

(9.60)

which is identical to the RG equations (9.26) and (9.27). Note that here the PAS signal increases as l' decreases or as a' decreases.

Finally, when l' becomes smaller than the fluid diffusion length ($\sigma'l' < 1$), we find

$$q = B'\alpha_t'\theta_0 = \frac{\gamma P_0}{T_0}\theta_0 e^{i\omega t} \qquad (9.61)$$

The PAS signal then becomes independent of both l' and a' and undergoes a $+\pi/4$ phase shift. This explains the roll-off observed in the PAS signal when the frequency becomes very low.

It should be noted that this analysis applies for a gas only at frequencies < 1000 Hz, since above this value, attenuation of the acoustic signal due to thermoviscous effects becomes important.

9.7   THERMALLY INDUCED VIBRATIONS OF THE SAMPLE

In the preceding sections of this chapter we consider the heating of the sample from the absorbed light, the diffusion of some of this heat to the surrounding gas or fluid, and the subsequent transport of the thermally induced pressure fluctuations in the gas. In this section we consider the contributions to the PAS signal from thermally induced mechanical vibrations of the sample itself. This problem has been analyzed by McDonald and Wetsel (1978). The analysis here, however, follows our previous treatment, which although simpler, provides results that are essentially identical to those obtained by McDonald and Wetsel.

Let us assume that the sample is a condensed fluid contained in a gas-microphone system. The periodic temperature $\theta(x, t)$ within the sample is given by (9.8) of the RG theory,

$$\theta(x, t) = (Ue^{\sigma x} + Ve^{-\sigma x} - Ee^{\beta x})e^{i\omega t} \qquad -l \leqslant x \leqslant 0 \qquad (9.62)$$

Let us further assume that the sample and backing have similar thermal properties, and that a gas is present in front of the sample. We then find

$$U \simeq \frac{\beta}{\sigma}E \qquad (9.63)$$

$$V \simeq 0 \qquad (9.64)$$

$$E = \frac{\beta I_0}{2\kappa(\beta^2 - \sigma^2)} \qquad (9.65)$$

Thus

$$\theta(x,t) \simeq (Ue^{\sigma x} - Ee^{\beta x})e^{i\omega t} \tag{9.66}$$

The general solution $u_0(x)$ for the acoustic wave equation is given by

$$u_0 = Ae^{ikx} + De^{-ikx} + \frac{\sigma \alpha_t U}{(k^2 + \sigma^2)}e^{\sigma x} - \frac{\beta \alpha_t E}{(k^2 + \beta^2)}e^{\beta x} \tag{9.67}$$

Note that x is always negative in the sample region.

Since a gas is on the $x = 0$ side and a rigid backing is on the $x = -l$ side, we can set the boundary conditions

$$p(0) = 0$$

$$u(-l) = 0 \tag{9.68}$$

which gives the following equations for A and D:

$$A = -\alpha_t U \left(\frac{1}{k^2 + \sigma^2} \right) \left(\frac{ik\sigma e^{-\sigma l} - k^2 e^{ikl}}{ik(e^{ikl} + e^{-ikl})} \right)$$

$$+ \alpha_t E \left(\frac{1}{k^2 + \beta^2} \right) \left(\frac{ik\beta e^{-\beta l} - k^2 e^{ikl}}{ik(e^{ikl} + e^{-ikl})} \right) \tag{9.69}$$

$$D = -\alpha_t U \left(\frac{1}{k^2 + \sigma^2} \right) \left(\frac{ik\sigma e^{-\sigma l} + k^2 e^{-ikl}}{ik(e^{ikl} + e^{-ikl})} \right)$$

$$+ \alpha_t E \left(\frac{1}{k^2 + \beta^2} \right) \left(\frac{ik\beta e^{-\beta l} + k^2 e^{-ikl}}{ik(e^{ikl} + e^{-ikl})} \right) \tag{9.70}$$

Thus the displacement at $x = 0$ is given by

$$u(0) = \left[A + D + \alpha_t U \left(\frac{\sigma}{k^2 + \sigma^2} \right) - \alpha_t E \left(\frac{\beta}{k^2 + \beta^2} \right) \right] e^{i\omega t}$$

$$= -\left\{ \alpha_t U \left(\frac{1}{k^2 + \sigma^2} \right) \left[\frac{2ik\sigma e^{-\sigma l} - k^2(e^{ikl} - e^{-ikl})}{ik(e^{ikl} + e^{-ikl})} \right] - \alpha_t U \left(\frac{\sigma}{k^2 + \sigma^2} \right) \right.$$

$$\left. - \alpha_t E \left(\frac{1}{k^2 + \beta^2} \right) \left[\frac{2ik\beta e^{-\beta l} - k^2(e^{ikl} - e^{-ikl})}{ik(e^{ikl} + e^{-ikl})} \right] + \alpha_t E \left(\frac{\beta}{k^2 + \beta^2} \right) \right\} e^{i\omega t} \tag{9.71}$$

There are resonance vibrations when $kl = (n+1)\pi/2$ or $l = (n+1)\lambda/4$.

At low frequencies when $kl \ll 1$ and when $k^2 < \sigma^2, \beta^2$

$$u(0) \simeq \left[\alpha_t U \frac{1}{\sigma^2}(\sigma e^{-\sigma l}) - \alpha_t U \frac{1}{\sigma} - \alpha_t E \frac{1}{\beta^2}(\beta e^{-\beta l}) + \alpha_t E \frac{1}{\beta} \right] e^{i\omega t}$$

$$= \left[\alpha_t U \left(\frac{1 - e^{-\sigma l}}{\sigma} \right) - \alpha_t E \left(\frac{1 - e^{-\beta l}}{\beta} \right) \right] e^{i\omega t} \tag{9.72}$$

For a thermally thick and non-opaque sample $\sigma l \gg 1$, $\sigma > \beta$

$$u(0) \simeq -\alpha_t E \left(\frac{1 - e^{-\beta l}}{\beta} \right) e^{i\omega t}$$

$$= -\frac{\alpha_t I_0}{2\kappa(\beta^2 - \sigma^2)}(1 - e^{-\beta l})e^{i\omega t} \tag{9.73}$$

or

$$u(0) = -\frac{i\alpha_t I_0}{2\rho C\omega}(1 - e^{-\beta l})e^{i\omega t} \tag{9.74}$$

This displacement $u(0)$ gives rise to an added pressure term at the microphone. Since the velocity of sound in a gas is much lower than that in fluids or solids, a frequency at which kl in the fluid or solid is $\ll 1$ may, however, be one for which $k'l'$ in the gas is not small. Thus, in general, the pressure at the microphone induced by the periodic displacement $u(0)$, is given by the solution of the homogeneous wave equation

$$\frac{\partial^2 u}{\partial x^2} + k'^2 u = 0 \tag{9.75}$$

with $u = u(0)$ as the boundary condition at $x = 0$, giving

$$u(x, t) = u(0) \left(\frac{e^{ik'l'}e^{-ik'x} - e^{-ik'l'}e^{ik'x}}{e^{ik'l'} - e^{-ik'l'}} \right) e^{i\omega t} \tag{9.76}$$

producing resonances whenever $l' = n\lambda'/2$.

The pressure on the microphone produced by this mechanical component is given by

$$q_m = -B' \frac{\partial u(l')}{\partial x}$$

$$= \gamma P_0 u(0) \left(\frac{2ik'}{e^{ik'l'} - e^{-ik'l'}} \right) e^{i\omega t} \tag{9.77}$$

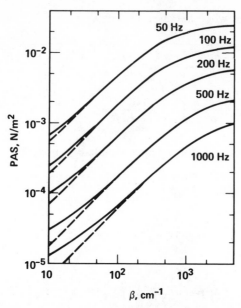

Figure 9.10   Theoretical dependence of the photoacoustic signal for an opaque aqueous dye solution. The solid lines are the results of (9.79), while the dashed lines are the results of the unmodified RG theory. (Reproduced by permission from McDonald and Wetsel, 1978.)

At low frequencies this simply becomes

$$q_m = \frac{\gamma P_0 u(0)}{l'} \qquad (9.78)$$

The total pressure on the microphone due both to thermal and mechanical effects at low frequencies for a thermally thick and non-opaque sample and for a thermally-thick air column is given by (9.74) and (9.40):

$$q_T = q + q_m$$

$$= \frac{-i\gamma P_0}{l'} \left(\frac{I_0}{2\rho C \omega} \right) \left[\frac{\beta\sigma}{\sigma' T_0 (\beta + \sigma)} + \alpha_t (1 - e^{-\beta l}) \right] e^{i\omega t} \qquad (9.79)$$

This is identical to expression (41) derived by McDonald and Wetsel (1978) using a composite-piston model. They found that this model accounted for their results equally as well as their more complex coupled model that required computer solution.

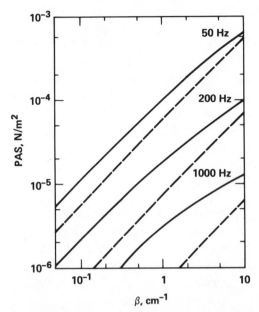

Figure 9.11 Theoretical dependence of the photoacoustic signal on absorption coefficient β for a weakly absorbing aqueous dye solution (Reproduced by permission from McDonald and Wetsel, 1978.)

Figures 9.10 and 9.11 show the predicted effects of the thermally induced vibrations of an aqueous dye solution sample as a function of both modulation frequency and absorption coefficient. The solid lines represent the PAS signal as determined from (9.79) above or (41) in McDonald and Wetsel's paper, and the dashed lines represent the PAS signal as determined from the RG theory. As the absorption coefficient is reduced, the effect of sample vibrations becomes appreciable, the relative effect increasing with modulation frequency. McDonald and Wetsel were able to confirm these predictions with experiments on aqueous solutions of phenol red sodium salts.

REFERENCES

Aamodt, L. C., Murphy, J. C., and Parker, J. G. (1977). *J. Appl. Phys.* **48**, 927.

Adams, M. J., Kirkbright, G. F., and Menon, K. R. (1979). *Anal. Chem.* **51**, 508.

Arnold, H. D., and Crandall, I. B. (1917). *Phys. Rev.* **10**, 22.

Bell, A. G. (1881). *Philos. Mag.* (5) **11**, 510.

Bennett, H. S., and Forman, R. A. (1976). *Appl. Opt.* **15**, 2405.

Hordvik, A. and Schlossberg, H. (1976). *Appl. Opt.* **16**, 101.

McClelland, J. F., and Kniseley, R. N. (1976). *Appl. Phys. Lett.* **28**, 467.

McDonald, F. A., and Wetsel, G. C. Jr. (1978). *J. Appl. Phys.* **49**, 2313.

Mercadier, M. E. (1881). *C. R. Hebd. Serv. Acad. Sci.* **92**, 409.

Monahan, E. M., Jr., and Nolle, A. W. (1977). *J. Appl. Phys.* **48**, 3519.

Parker, J. G. (1973). *Appl. Opt.* **12**, 2974.

Preece, W. H. (1881). *Proc. R. Soc. Lond.* **31**, 506.

Rayleigh (Lord) (1881). *Nature (Lond.)* **23**, 274.

Rosencwaig, A. (1973). *Opt. Commun.* **7**, 305.

Rosencwaig, A. (1977). *Rev. Sci. Instrum.* **48**, 1133.

Rosencwaig, A. (1978). In *Advances in Electronics and Electron Physics*, Vol. 46 (L. Marton, Ed.), pp. 207–311, Academic Press, New York.

Rosencwaig, A., and Gersho, A. (1975). *Science* **190**, 556.

Rosencwaig, A., and Gersho, A. (1976). *J. Appl. Phys.* **47**, 64.

Wetsel, G. C., Jr., and McDonald, F. A. (1977a). *Appl. Phys. Lett.*, **30**, 252.

Wetsel, G. C., Jr., and McDonald, F. A. (1977b). *Bull. Am. Phys. Soc.*, **22**, 295.

White, R. M. (1963). *J. Appl. Phys.*, **34**, 3559.

10

GENERAL THEORY OF THE PHOTOACOUSTIC EFFECT IN CONDENSED MEDIA: THE PIEZOELECTRIC SIGNAL

10.1 INTRODUCTION

The photoacoustic effect arises when intensity-modulated light, or some other form of electromagnetic radiation, impinges on a sample. If the sample absorbs any of the incident radiation, energy levels within the sample medium are excited, and when these levels deexcite, they generally do so through nonradiative or heat-producing processes. Thus the absorption of intensity-modulated electromagnetic radiation at any point in the sample results in a periodic localized heating of the sample medium.

When localized heating occurs in a material, the heat energy is transmitted to the surrounding matter through two mechanisms. First, there is a diffusion of heat from the originally heated area to the surrounding area by means of heat conduction and diffusion. The rate of energy transfer by this means is determined by the material's thermal diffusivity ($\alpha = \kappa / \rho C$, where κ is the thermal conductivity, ρ is the density, and C is the specific heat). When the heating is periodic at a frequency ω, then the distance of appreciable transfer of a periodic heat signal through the medium is given by the thermal diffusion length $\mu, [\mu = (2\alpha / \omega)^{1/2}]$. Energy transfer through thermal diffusion is a dissipative process, in which individual atoms, ions, or molecules within the material are vibrationally excited in a noncooperative manner. We refer to this mode of energy transfer as the thermoacoustic mode.

The second mode of energy transfer is through a coupling of the local heat energy to the vibrational modes of the material itself, that is, through a coupling to the sample's acoustic phonon spectrum. This is a thermoelastic process, which is generally nondissipative. The speed of this energy transfer is governed by the speed of sound in the material, and the distance of appreciable energy transfer is limited solely by the dimensions of the sample or by some other boundary condition, except at very high frequencies where ultrasonic attenuation can occur.

10.2 THERMOELASTIC THEORY

The first evidence for the generation of elastic (acoustic) waves in condensed media from the absorption of electromagnetic radiation was obtained in experiments reported by Michaels (1961) and White (1962; 1963a). These experiments used high-powered light sources such as electric arcs and pulsed ruby lasers. In a classic paper referred to earlier, White (1963b) developed the theory for this effect, treating elastic wave generation in isotropic elastic bodies under several transient heating conditions. In this chapter, we continue the analysis of the previous chapter to compute those stresses produced within a photoacoustically probed sample that can be detected with a piezoelectric device.

For an isotropic solid, (9.43) must be modified so that the stress–strain relation now becomes

$$p(x) = (\lambda^* + 2\mu^*)\varepsilon(x) - B\alpha_t \theta(x) \tag{10.1}$$

where λ^* and μ^* are, respectively, the first Lamé constant and the modulus of rigidity. Furthermore, the sound velocity is given by

$$c_0^2 = \frac{1}{\rho}(\lambda^* + 2\mu^*) \tag{10.2}$$

The equation of motion for particle displacement then becomes

$$\frac{\partial^2 u}{\partial x^2} + k^2 u = \eta \alpha_t \frac{\partial \theta}{\partial x} \tag{10.3}$$

where

$$\eta = \frac{B}{\rho c_0^2} \tag{10.4}$$

Let us now evaluate the stresses and strains produced in a photoacoustic sample. As in the preceding chapter, we assume that the sample is thermally thick and thus that the spatial temperature profile is given by

$$\theta(x) = Ue^{-\sigma x} - Ee^{-\beta x} \tag{10.5}$$

where we assume x is now positive and

$$0 \leqslant x \leqslant l \tag{10.6}$$

The general solution for $u(x)$ is

$$u(x) = Ae^{-ikx} + De^{ikx} - \eta\alpha_t U\left(\frac{\sigma}{k^2+\sigma^2}\right)e^{-\sigma x} + \eta\alpha_t E\left(\frac{\beta}{k^2+\beta^2}\right)e^{-\beta x}$$

$$(10.7)$$

and

$$\frac{\partial u}{\partial x} = -ikAe^{-ikx} + ikDe^{ikx} + \eta\alpha_t U\left(\frac{\sigma^2}{k^2+\sigma^2}\right)e^{-\sigma x} - \eta\alpha_t E\left(\frac{\beta^2}{k^2+\beta^2}\right)e^{-\beta x}$$

$$(10.8)$$

10.3 UNCONSTRAINED CASE

Most piezoelectric experiments are performed with the sample free to expand and contract and with the transducer mounted so as to minimally impede this motion. We can, therefore, assume that the boundary conditions for such a case are unconstrained ends at $x=0$ and $x=l$:

$$p(0) = 0$$
$$p(l) = 0$$

$$(10.9)$$

Then

$$p(x) = \rho c_0^2 \frac{\partial u}{\partial x} - B\alpha_t \theta(x) \qquad (10.10)$$

Therefore

$$\frac{p(x)}{\rho c_0^2} = \frac{\partial u}{\partial x} - \eta\alpha_t\theta(x) \qquad (10.11)$$

Applying the boundary conditions of (10.9) gives the following expressions for A and D:

$$ikA = \eta\alpha_t U\left(\frac{k^2}{k^2+\sigma^2}\right)\left(\frac{e^{-\sigma l}-e^{ikl}}{e^{ikl}-e^{-ikl}}\right) - \eta\alpha_t E\left(\frac{k^2}{k^2+\beta^2}\right)\left(\frac{e^{-\beta l}-e^{ikl}}{e^{ikl}-e^{-ikl}}\right)$$

$$(10.12)$$

$$ikD = \eta\alpha_t U\left(\frac{k^2}{k^2+\sigma^2}\right)\left(\frac{e^{-\sigma l}-e^{-ikl}}{e^{ikl}-e^{-ikl}}\right) - \eta\alpha_t E\left(\frac{k^2}{k^2+\beta^2}\right)\left(\frac{e^{-\beta l}-e^{-ikl}}{e^{ikl}-e^{-ikl}}\right)$$

$$(10.13)$$

We then get

$$\frac{p(x)}{\rho c_0^2} = \eta\alpha_t U\left(\frac{k^2}{k^2+\sigma^2}\right)\left[\frac{(e^{ikl}-e^{-\sigma l})e^{-ikx}-(e^{-ikl}-e^{-\sigma l})e^{ikx}}{e^{ikl}-e^{-ikl}}\right]$$

$$-\eta\alpha_t E\left(\frac{k^2}{k^2+\beta^2}\right)\left[\frac{(e^{ikl}-e^{-\beta l})e^{-ikx}-(e^{-ikl}-e^{-\beta l})e^{ikx}}{e^{ikl}-e^{-ikl}}\right]$$

$$-\eta\alpha_t U\left(\frac{k^2}{k^2+\sigma^2}\right)e^{-\sigma x}+\eta\alpha_t E\left(\frac{k^2}{k^2+\beta^2}\right)e^{-\beta x} \qquad (10.14)$$

For an optically opaque and thermally thick sample of large l,

$$\frac{p(x)}{\rho c_0^2} \simeq \eta\alpha_t U\left(\frac{k^2}{k^2+\sigma^2}\right)(e^{-ikx}-e^{-\sigma x})$$

$$-\eta\alpha_t E\left(\frac{k^2}{k^2+\beta^2}\right)(e^{-ikx}-e^{-\beta x}) \qquad (10.15)$$

For $\beta\gg\sigma$, as in the case White considered, $U\gg E$ and

$$\frac{p(x)}{\rho c_0^2} \simeq \eta\alpha_t U\left(\frac{k^2}{k^2+\sigma^2}\right)(e^{-ikx}-e^{-\sigma x})$$

$$= \eta\alpha_t\theta_0\left(\frac{k^2}{k^2+\sigma^2}\right)(e^{-ikx}-e^{-\sigma x}) \qquad (10.16)$$

where

$$\theta_0 = \frac{(1-i)I_0}{2\sqrt{2}\,(\rho C\kappa\omega)^{1/2}} = \text{surface temperature} \qquad (10.17)$$

Of course, at $x=0$ and $x=l$, $p(x)=0$. Also, at low and moderate frequencies where $kl\ll 1$,

$$\frac{p(x)}{\rho c_0^2} \simeq \eta\alpha_t U\left(\frac{k^2}{k^2+\sigma^2}\right)\left(\frac{l-x(1-e^{-\sigma l})}{l}\right)$$

$$-\eta\alpha_t E\left(\frac{k^2}{k^2+\beta^2}\right)\left(\frac{l-x(1-e^{-\beta l})}{l}\right)-\eta\alpha_t U\left(\frac{k^2}{k^2+\sigma^2}\right)e^{-\sigma x}$$

$$+\eta\alpha_t E\left(\frac{k^2}{k^2+\beta^2}\right)e^{-\beta x} \qquad (10.18)$$

We find that as ω approaches zero, $p(x) \to 0$ for all x.
This is, of course, reasonable, since a completely free body in which the acoustic wavelength is larger than the geometric optical and thermal lengths experiences no internal stresses. Such a body simply experiences a net displacement, and strains exist only in the regions where temperature changes occur.

10.4 CONSTRAINED CASE

The usual piezoelectric arrangement when working with fluids is to constrain the fluid within a piezoelectric cell. For such cases, the boundary conditions in one dimension are

$$u(0) = 0 \qquad u(l) = 0 \tag{10.19}$$

We find

$$A = \eta \alpha_t U \left(\frac{\sigma}{k^2 + \sigma^2} \right) \left(\frac{e^{ikl} - e^{-\sigma l}}{e^{ikl} - e^{-ikl}} \right) - \eta \alpha_t E \left(\frac{\beta}{k^2 + \beta^2} \right) \left(\frac{e^{ikl} - e^{-\beta l}}{e^{ikl} - e^{-ikl}} \right) \tag{10.20}$$

$$D = -\eta \alpha_t U \left(\frac{\sigma}{k^2 + \sigma^2} \right) \left(\frac{e^{-ikl} - e^{-\sigma l}}{e^{ikl} - e^{-ikl}} \right) + \eta \alpha_t E \left(\frac{\beta}{k^2 + \beta^2} \right) \left(\frac{e^{-ikl} - e^{-\beta l}}{e^{ikl} - e^{-ikl}} \right) \tag{10.21}$$

We then obtain

$$\frac{p(x)}{\rho c_0^2} = -\eta \alpha_t U \left(\frac{ik\sigma}{k^2 + \sigma^2} \right) \left(\frac{(e^{ikl} - e^{-\sigma l})e^{-ikx} + (e^{-ikl} - e^{-\sigma l})e^{ikx}}{e^{ikl} - e^{-ikl}} \right)$$

$$+ \eta \alpha_t E \left(\frac{ik\beta}{k^2 + \beta^2} \right) \left(\frac{(e^{ikl} - e^{-\beta l})e^{-ikx} + (e^{-ikl} - e^{-\beta l})e^{ikx}}{e^{ikl} - e^{-ikl}} \right)$$

$$- \eta \alpha_t U \left(\frac{k^2}{k^2 + \sigma^2} \right) e^{-\sigma x} + \eta \alpha_t E \left(\frac{k^2}{k^2 + \beta^2} \right) e^{-\beta x} \tag{10.22}$$

For $kl \ll 1$

$$\frac{p(x)}{\rho c_0^2} \simeq -\eta \alpha_t U\left(\frac{\sigma}{k^2+\sigma^2}\right)\left(\frac{1-e^{-\sigma l}}{l}\right) + \eta \alpha_t E\left(\frac{\beta}{k^2+\beta^2}\right)\left(\frac{1-e^{-\beta l}}{l}\right)$$

$$-\eta \alpha_t U\left(\frac{k^2}{k^2+\sigma^2}\right)e^{-\sigma x} + \eta \alpha_t E\left(\frac{k^2}{k^2+\beta^2}\right)e^{-\beta x} \qquad (10.23)$$

And for $k^2 < \sigma^2, \beta^2$

$$\frac{p(x)}{\rho c_0^2} \simeq -\eta \alpha_t U\left(\frac{1-e^{-\sigma l}}{\sigma l}\right) + \eta \alpha_t E\left(\frac{1-e^{-\beta l}}{\beta l}\right) \qquad (10.24)$$

But the right-hand side of (10.24) is simply the stress due to the average temperature rise of the sample since

$$\bar{\theta} = \frac{1}{l}\int_0^l \theta(x)\,dx = U\left(\frac{1-e^{-\sigma l}}{\sigma l}\right) - E\left(\frac{1-e^{-\beta l}}{\beta l}\right) \qquad (10.25)$$

Therefore

$$p(x) = B\alpha_t \bar{\theta} \qquad (10.26)$$

Thus for the constrained fluid $p(x)$ is independent of x and varies as ω^{-1} at low frequencies. The magnitude of the PAS signal is simply dependent on the total amount of light absorbed by the sample.

Experiments on a constrained fluid have been performed by Lahmann et al. (1977) and by Oda et al. (1978). Both groups found that the photoacoustic signal does indeed vary as ω^{-1} for frequencies at least up to 1 kHz.

10.5   THREE-DIMENSIONAL CASE

The calculations in this chapter are all for the one-dimensional case. This one-dimensional treatment is almost never appropriate for a real solid sample, although uniform illumination of one complete surface of an unconstrained cylindrical sample with isotropic elastic properties might approximate such a situation. In general, photoacoustic stress calculations have to be performed for three-dimensional samples with boundaries, and in which shear stresses, as well as compressional or longitudinal stresses, also have to be considered. Such calculations are quite complex, and few can be carried out analytically as Nowacki (1962) has clearly demonstrated for the analogous problems of thermoelasticity. One of the major differences between the one-dimensional case (fluid) and the three-dimensional case (solid) is that stresses and strains exist in the case of the solid in regions of the sample where there is no change in local temperature, whereas, as we see above, such stresses vanish in the unconstrained fluid.

10.6 CONCLUSIONS

Piezoelectric detection of a photoacoustic signal offers considerable advantages over the gas-microphone method for certain types of samples. For example, a large sample would be difficult to accommodate within a gas-microphone system, and if a PAS cell were made large enough to accommodate it, the signal obtained would be quite low. Another type of sample for which the piezoelectric method is better suited is the bulk sample that has low to moderate optical absorption. In a gas-microphone system, the signal would be proportional only to the light absorbed within a thermal diffusion length of the surface. For such samples, this would be only a very small part of the total light absorbed. On the other hand, piezoelectric detection can be made sensitive to the total amount of light absorbed by the entire sample, thereby providing a more efficient measure of the absorbed energy.

In piezoelectric detection we no longer need to consider the properties of the surrounding gas; in fact, the sample could be in a vacuum. Still, we must contend with new complications. For example, as we show in this chapter, the strength and phase of the PAS signal are dependent on the boundary conditions acting on the sample, and on how and where we mount the piezoelectric detector.

Finally, the piezoelectric method offers a distinct advantage in the time domain. A gas-microphone system is usually limited by the relatively low-frequency response of the microphone (50 kHz). A piezoelectric detector can operate at much higher frequencies, and in fact, can go as high as 1 GHz in some situations.

REFERENCES

Lahmann, W., Ludewig, H. J., and Welling, H. (1977). *Chem. Phys. Lett.* **45**, 177.

Michaels, J. E. (1961). *Planetary Space Sci.* **7**, 427.

Nowacki, W. (1962). *Thermoelasticity*, Addison-Wesley, Reading, Massachusetts.

Oda, S., Sawada, T., and Kamada, H. (1978). *Anal. Chem.* **50**, 865.

White, R. M. (1962). *IEEE Trans. Instrum.* **I–II**, 294.

White, R. M. (1963a). *J. Appl. Phys.* **34**, 2123.

White, R. M. (1963b). *J. Appl. Phys.* **34**, 3559.

PHOTOACOUSTIC THEORY MADE EASY

11.1 INTRODUCTION

The preceding chapters demonstrate that the mathematical analysis of the photoacoustic signal, whether measured by a gas-microphone or by a piezoelectric method, is fairly laborious and complex. It is perfectly understandable, therefore, that many researchers in the field are somewhat dismayed by the apparent difficulty of performing quantitative analysis of photoacoustic data. We hope to show in this chapter, that this dismay is not justified, and that, in most cases, an analysis can be performed using quite simple theoretical expressions that can be derived from basic physical insights.

11.2 GAS-MICROPHONE SIGNAL

For a gas-microphone photoacoustic system, the pressure or stress in the gas (or in any fluid) that is generated within a thermal diffusion length of the sample surface can be approximated by

$$p_{\mu'} \simeq B\alpha'_t\left(\tfrac{1}{2}\theta_0\right) \tag{11.1}$$

where $\tfrac{1}{2}\theta_0$ is approximated as the average temperature within this thermal diffusion length if θ_0 is the temperature at the sample–fluid interface. The pressure a distance l' away is given by

$$p = p_{\mu'}\left(\frac{\mu'}{l'}\right) = \tfrac{1}{2}B\alpha'_t\theta_0\left(\frac{\mu'}{l'}\right)$$

$$= \frac{\gamma P_0}{2T_0 a'l'}\theta_0 \tag{11.2}$$

for a gas. When the gas or fluid is completely constrained at its borders, then the pressure p is the same everywhere in the cell, as long as cell dimensions are much smaller than the acoustic wavelength.

The temperature θ_0 at the sample–gas interface can be approximated by

$$\theta_0 \simeq \frac{H_{abs}}{M_{th}} \tag{11.3}$$

where H_{abs} is the amount of heat absorbed per unit time within the first thermal diffusion length in the sample, and M_{th} is the thermal mass of this region of the sample. For the case where the thermal diffusion length is smaller than the sample thickness ($\mu < l$), and if A is the area illuminated by the light,

$$H_{abs} \simeq \frac{I_0 A(1-e^{-\beta\mu})}{\omega} \tag{11.4}$$

and

$$M_{th} = \rho C \mu A \tag{11.5}$$

This then gives

$$\theta_0 = \frac{I_0(1-e^{-\beta\mu})}{\rho C \omega \mu} \tag{11.6}$$

Similarly, when $\mu > l$, then

$$H_{abs} \simeq \frac{I_0 A(1-e^{-\beta l})}{\omega} \tag{11.7}$$

$$M_{th} \simeq \rho''C''\mu''A \tag{11.8}$$

and

$$\theta_0 = \frac{I_0(1-e^{-\beta l})}{\rho''C''\omega\mu''} \tag{11.9}$$

where the unprimed symbols represent the sample and the double-primed symbols represent the backing parameters.

Thermal conduction introduces a phase lag in the PAS signal. There is a $\pi/4$ phase lag due to conduction in the gas and an additional phase lag due to conduction in the sample. The sample phase lag is approximated by $\phi = 1/\beta\mu$ for $\beta^{-1} < \mu$. When $\beta^{-1} > \mu$, ϕ reaches a maximum value of $\pi/4$ as well.

Combining our expressions for the magnitude of the PAS signal with the above remarks on phase, we can reconstruct all six of the special RG cases considered in Chapter 9. We find that aside from a factor of $1/\sqrt{2}$, all six cases are properly given by the above arguments.

11.3   THERMALLY THIN GAS LENGTH

The roll-over in the PAS signal when the gas length becomes thermally thin ($l' < \mu'$) can be approximated by noting that if the front window of the cell is maintained at $\theta = 0$ relative to the sample, then the effective average temperature experienced by the gas is given by

$$\bar{\theta}' \simeq \tfrac{1}{2}\theta_0(1 - e^{-l'/\mu'}) \tag{11.10}$$

and

$$p' = \frac{\gamma P_0}{2T_0 a'}\left(\frac{1 - e^{-l'/\mu'}}{l'}\right)\theta_0 \tag{11.11}$$

which rolls over and becomes constant at $l'/\mu' \simeq 1.4$, in agreement with the results of Chapter 9 and those of Aamodt et al. (1977).

11.4   SAMPLE VIBRATIONS

We can approximate sample vibrations for a thermally thick sample by noting that when the front surface of the sample is free and the back surface is constrained, the front surface expands a distance equal to the total expansion of the sample; that is,

$$u_m(0) \simeq \alpha_t \frac{H_{\text{abs}}(l)}{M_{\text{th}}}l$$

$$= \alpha_t \frac{I_0(1 - e^{-\beta l})}{\rho C \omega} \tag{11.12}$$

which, aside from a factor of 2, agrees with (9.74) of Chapter 9. This displacement produces a pressure in the gas given by

$$p_m \simeq \gamma P_0 \frac{u_m(0)}{l'} \tag{11.13}$$

in agreement with the results of Chapter 9.

11.5 PIEZOELECTRIC SIGNALS

The photoacoustic piezoelectric signal can also be derived in a similar manner, by noting that in the near-field, low-frequency case, the photoacoustic stress in a constrained sample is

$$p = B\alpha_t \bar{\theta}$$

$$= B\alpha_t \frac{I_0(1 - e^{-\beta l})}{\rho C \omega l} \tag{11.14}$$

where again $\bar{\theta}$ represents the average temperature rise of the entire sample.

At high frequencies, where we are dealing with the far-field case, we can use a variation of White's simple formula (White, 1963).

$$p \simeq B\alpha_t \theta_0 \left(\frac{k^2}{k^2 + \chi^2} \right) \tag{11.15}$$

where the surface temperature θ_0 is given by (11.6) and (11.9) above, k ($= \omega/c_0$) is the acoustic wave vector, and χ is σ when $\sigma > \beta$, and $\chi = \beta$ when $\sigma < \beta$. The term $k^2/(k^2 + \chi^2)$ can be regarded as giving the extent of coupling between the heat-absorbing region and the acoustic wave.

11.6 FURTHER THEORETICAL DEVELOPMENTS

Although the photoacoustic theory is fairly well developed at this stage, there are several aspects that still remain to be analyzed.

11.6.1 Sample Surface Area

The present theory deals exclusively with samples of simple geometrical shapes and planar surfaces. It is well known, experimentally, that a finely ground powder gives a stronger photoacoustic signal than the same sample in an equally thin but integral form. Clearly, the PAS signal increases with an increase in the surface/volume ratio, on the assumption that the total heat transferred to the gas is proportional to this ratio. This problem should, however, be solved explicitly, particularly since PAS experiments on powder samples are of considerable importance.

11.6.2   Refinements of the Piezoelectric Theory

In Chapter 10 we deal with a simple isotropic medium in one dimension. A more complete treatment of the photoacoustic signal in a solid sample must consider three dimensions and take into account the effects of shear waves in addition to the compressional waves, the effects of sample boundaries, the effects of surface waves, and the effects that arise when the illuminated area is much smaller than the actual sample area.

REFERENCES

Aamodt, L. C., Murphy, J. C., and Parker, J. G. (1977). *J. Appl. Phys.* **48**, 927.

White, R. M. (1963). *J. Appl. Phys.* **34**, 3559.

PHOTOACOUSTIC SPECTROMETERS FOR CONDENSED SAMPLES

12.1 INTRODUCTION

In Chapter 4, we describe the design of a photoacoustic spectrometer used for the study of gaseous samples. A PAS spectrometer for condensed samples is quite similar, in that it too consists of a source of intensity-modulated optical radiation, a monochromator if necessary, a photoacoustic cell, and the electronics for detecting and storing the photoacoustic signal. The significant difference between the two spectrometers is in the design of the photoacoustic cell. In the first part of this chapter we discuss the design of a gas-microphone cell, and in the second part we consider a photoacoustic cell utilizing a piezoelectric detector.

12.2 THE GAS-MICROPHONE CELL

The experimental chamber or cell for a gas-microphone PAS system is the section containing the sample and microphone, with its preamplifier. Both a conventional condenser microphone with external biasing and an electret microphone with internal self-biasing provided from a charged electret foil have been found suitable.

Some criteria governing the actual design of the photoacoustic cell are:

1. Acoustic isolation from the outside world.
2. Minimization of extraneous photoacoustic signal arising from the interaction of the light beam with the walls, the windows, and the microphone in the cell.
3. Microphone configuration.
4. Means for maximizing the acoustic signal within the cell.
5. The requirements set by the samples to be studied and the type of experiments to be performed.

Let us now consider these criteria in more detail.

1.   We have found that the problem of acoustic isolation is not particularly serious providing one uses lock-in detection methods for analyzing the microphone signal. One should, of course, use chopping frequencies different from those present in the acoustic and vibrational spectrum of the environment. In addition one should design the cell with good acoustic seals and with walls of sufficient thickness to form a good acoustic barrier. Some reasonable precautions to isolate the photoacoustic cell from room vibrations should also be taken.

2.   To minimize any photoacoustic signal that may arise from the interaction of the light beam with the walls and windows of the cell, one should employ windows as optically transparent as possible for the wavelength region of interest and construct the body of the cell out of polished aluminum or stainless steel. Although the aluminum or stainless steel walls absorb some of the incident and scattered radiation, the resultant photoacoustic signal will be quite weak, as long as the thermal mass of these walls is large. A large thermal mass results in a small temperature rise at the surface and thus a small photoacoustic signal. In addition, one should keep all inside surfaces clean to minimize photoacoustic signals from surface contaminants. One should also design the cell so as to minimize the amount of scattered light that can reach the microphone diaphragm.

3.   Various microphone configurations can be used. We have found that both cylindrical microphones (Figure 12.1) and flat microphones (Figure 12.2) can readily be used. Cylindrical microphones have the

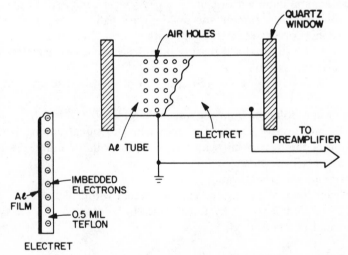

Figure 12.1 A simple photoacoustic cell employing a cylindrical electret microphone. (Reproduced by permission from Rosencwaig, 1977.)

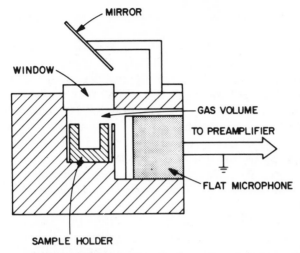

Figure 12.2 A simple photoacoustic cell employing a flat microphone. (Reproduced by permission from Rosencwaig, 1977.)

advantage of being easy to construct and have a large surface area, which increases their sensitivity. A disadvantage is that they usually do not possess a flat frequency response over a large acoustic range. This can be troublesome if one is planning to do experiments at different chopping frequencies. Flat microphones are commercially available, are quite sensitive when of reasonable size ($\frac{1}{2}$ in. diameter or larger) and good quality, and flat microphones possess a flat frequency response over a wide acoustic range. A good example of such a microphone is the General Radio Electret Microphone, Model 1961–9601.

The microphone chamber used for the cell depicted in Figure 12.1 is 1 in. in diameter and 5 in. long. The electret foil is 0.001 in. Teflon sheet with a 1000 Å aluminum coating on one side. The charge density was ∼10 nC/cm^2. These foils were supplied by J. E. West of Bell Laboratories. The cylindrical chamber was machined to have a large number of closely spaced holes, and small circumferential groves ∼0.001 in. high on the outside wall upon which the electret foil rested. Because of the rather excessive volume of the cell, the signal away from mechanical resonance of the electret foil was quite low. At resonance, however (∼400 Hz), the signal was quite high, giving an output of 1 mV at 500 nm when a carbon black absorber was used in the cell. The Q at resonance was measured to be ∼30.

The electret microphone used in the cell of Figure 12.2 was made at Bell Laboratories according to the design specified by Sessler and West (1966).

These microphones have a rated sensitivity of ∼1 mV/μbar, with a flat frequency response from 50 Hz to 15 KHz.

For both the cylindrical and flat microphone, we used an Ithaco low-noise 40 dB preamplifier (Model 144). The signal from the preamplifier was then fed into a Princeton Applied Research lock-in amplifier (Model 129A). The cell shown in Figure 12.2 had a volume of ∼5 cm³. With a carbon-black absorber, the signal at 100 Hz and 550 nm for the incident light was ∼1 mV. The signal/noise ratio for the carbon-black absorber was ∼1000 : 1. A study of the noise indicated that it was mainly electronic noise, primarily from the microphone–preamplifier system. This noise level is several orders of magnitude greater than the photoacoustic theoretical limit imposed by Brownian motion (Kreuzer, 1971; see also Chapter 4).

4. Since the signal in a photoacoustic cell used for solid samples varies inversely with the gas volume as shown in (9.31), one should attempt to minimize the gas volume. However, one must take care not to minimize this volume to the point that the acoustic signal produced at the sample suffers appreciable dissipation to the cell window and walls before reaching the microphone.

The distance between the sample and the cell window should always be greater than the thermal diffusion length of the gas, since as shown in Chapter 9, it is this boundary layer of gas that acts as an acoustic piston generating the signal in the cell. When the column of gas l' in front of the sample becomes comparable to the gaseous thermal diffusion length μ', then the photoacoustic signal decreases as shown in Chapter 9.

Thus it is clear that one should maintain $l' > \mu'$ for all chopping frequencies of interest. Since $\mu' \propto (\omega)^{-1/2}$, we must consider the lowest chopping frequencies that we are likely to employ. For air at room temperature and pressure, the thermal diffusion length is ∼0.06 cm at a chopping frequency of 10 Hz. A reasonable value for l' would then be about 0.1 cm.

In the design of a suitable photoacoustic cell, one must take thermoviscous damping into account as well, since this could be a source of significant signal dissipation to the cell boundaries. Thermoviscous damping results in a $e^{-\varepsilon x}$ attenuation, where ε is a damping coefficient given by (Kinsler and Frey, 1962),

$$\varepsilon = \frac{1}{dc_0}\left(\frac{\eta_e \omega}{2\rho_0}\right)^{1/2} \tag{12.1}$$

where d is the closest dimension between cell boundaries, as in the passageway, c_0 is the sound velocity, ω is the frequency, ρ_0 is the gas density, and η_e is an effective viscosity, which is dependent on both the ordinary viscosity and thermal conductivity of the gas. Again, for air at room temperature and pressure, there is negligible thermoviscous damping at 100 Hz as long as $d > 0.01$ cm. It should be noted, however, that whereas the thermal diffusion length varies as $(\omega)^{-1/2}$, the thermoviscous damping coefficient varies as $(\omega)^{1/2}$. Thus, while the thermal diffusion length is the predominant parameter at low frequencies, the thermovicous term is predominant at high frequencies. A cell designed to be used over a wide range of frequencies while being capable of handling a number of different gases should then have a minimum distance between the sample and window and minimum passageway dimensions of 1–2 mm.

The acoustic signal in the photoacoustic cell can be further enhanced by a number of means. For example, if it is possible to operate at only one frequency, one can take advantage of the nonflat frequency response of a cylindrical microphone and work at the frequency of its peak sensitivity. Another means is by making use of an acoustically resonant cell section with a length equal to $n\lambda/2$, where n is an integer and λ is the acoustical wavelength (Chapter 9). To achieve such a resonance effect without making the cell gas volume too large, one can construct a cell similar to that shown in Figure 12.3. Here the sample is placed in a nonresonant section of the cell, with a resonant section of reasonably small cross-sectional area joining the sample section with the microphone. One can, in addition, place a suitable diaphragm between the nonresonant and resonant sections, and between the resonant section and the microphone,

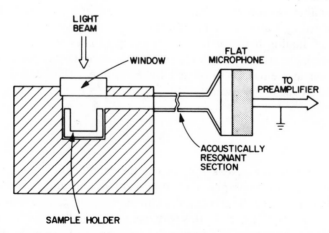

Figure 12.3 A photoacoustic cell with an acoustically resonant section. (Reproduced by permission from Rosencwaig, 1977.)

thereby allowing, for example, different gases or different gas pressures to be used in the separate sections. In the use of resonant cell designs one, however, is limited to a fixed chopping frequency.

Other methods to enhance the acoustic signal include the use of gases with a higher thermal conductivity and the use of higher gas pressures and of lower gas temperatures. The theory shows that the photoacoustic signal varies in most cases as $[(\kappa'p_0)^{1/2}]/T_0$, where κ' is the gas thermal conductivity, P_0 is the pressure, and T_0 is the temperature of the gas. All these methods increase the photoacoustic signal without limiting the choice of chopping frequencies.

5.   How closely one adheres to the above criteria, of course, depends on the type of sample used (powder, smear, liquid, etc.), its size, and the type of experiment one wishes to perform (low temperature, high temperature, etc.).

12.3   SIMPLE CELLS

There have been several simple cell designs proposed by various investigators. For example, Figure 12.4 depicts a cell proposed by Gray et al. (1977). This cell consists of a Plexiglass housing with a stainless steel chamber liner. The microphone used is an inexpensive Knowles Laboratories Model BW-1789 electret microphone. The sample is mounted on the removable aluminum sample probe F and it and the window are sealed with rubber O-rings. Another very simple cell was designed by Ferrell and Haven (1977).

Still another version of a simple cell is shown in Figure 12.5. This cell was designed by Cahen et al. (1978). It is particularly suited for very small samples of liquids that are held in the cup (*1*). In this cell, it is also possible to make use of Helmholtz resonance since the sample and microphone compartments are separated by a channel. The basic Helmholtz resonance frequency is given by

$$\nu_0 = \frac{c_0}{2}\left(\frac{\pi r^2}{dV_e}\right) \tag{12.2}$$

where r is the channel radius, d is the channel length, and $V_e = V_m V_s/(V_m + V_s)$, where V_m is the microphone compartment volume and V_s is the sample compartment volume. By moving the plunger (*3*), the sample compartment volume V_s can be changed. In this way one can partially circumvent one of the main disadvantages of acoustically resonant cells, namely, that they can be used only at one fixed frequency.

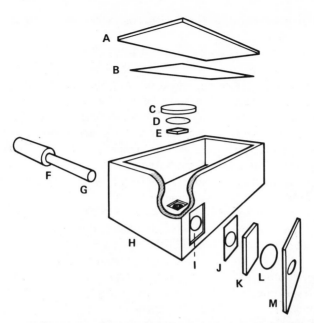

Figure 12.4 Construction details of a simple photoacoustic cell. (*A*) Cell lid (Plexiglas), (*B*) follower preamplifier circuit board, (*C*) microphone part pressure seal (Plexiglas), (*D*) rubber 0-ring pressure seal, (*E*) microphone (*F*) sample probe (aluminum), (*G*) rubber 0-ring pressure seal, (*H*) cell housing (Plexiglas), (*I*) sample chamber liner (stainless steel), (*J*) Teflon pressure seal gasket, (*K*) quartz window, (*L*) rubber 0-ring compression gasket, (*M*) window assembly hold-down (aluminum). (Reproduced by permission from Gray et al., 1977.)

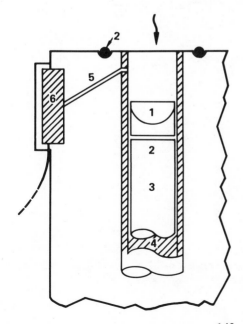

Figure 12.5 Top part of a simple photoacoustic cell (optical window not shown). (*1*) Sample cup (1.5 mm deep, 2 mm radius), (*2*) 0-ring, (*3*) plunger, (*4*) optional metal liner, (*5*) channel, (*6*) microphone compartment. The plunger can be moved, taking the top of the sample cup from 4 to 19 mm from the top of the cell. The channel to the microphone is 6 mm long and has 0.5-mm radius. (Reproduced by permission from Cahen et al., 1978.)

143

12.4   DATA ACQUISITION

The tasks of acquiring, storing, and displaying the data can be performed in many ways. However, certain basic procedures should be followed. For example, the signal from the microphone preamplifier should be processed by an amplifier tuned to the chopping frequency to maximize the signal/noise ratio. If phase, as well as signal amplitude, is desired, or if very weak signals are to be measured, then a phase-sensitive lock-in amplifier should be used. Figure 12.6 shows a block diagram of a typical single-beam photoacoustic spectrometer.

For a single-beam spectrometer, provisions must generally be made to remove, from the photoacoustic spectrum, any spectral structure due to the lamp, monochromator, and optics of the system. This normalization can be conveniently done by digitizing the analog signal from the tuned amplifier and then performing a point-by-point normalization (i.e., division) with either a power meter reading or a previously recorded photoacoustic spectrum obtained with a black absorber.

Although commercial data acquisition systems are fairly expensive, it is possible, as Eaton and Stuart (1978) have shown, to put together a fairly

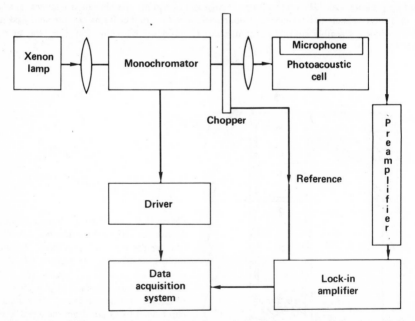

Figure 12.6   Block diagram of a typical single-beam photoacoustic spectrometer utilizing a gas-microphone cell. (Reproduced by permission from Rosencwaig, 1977.)

inexpensive system using commonly available and easily constructed components. ·

12.5   CALIBRATION–A PHOTOACOUSTIC THERMOPHONE

In a gas–microphone PAS system, the photoacoustic signal is dependent on various parameters of cell design, gas, backing material, microphone, and so on. Since the basic objective of a PAS experiment is usually the determination of the sample properties, these cell-dependent effects are experimental impediments. Fortunately, for fixed-modulation frequency experiments, these complications do not interfere with wavelength-dependent measurements of the sample. However, for experiments involving variations in modulation frequency, or for direct comparisons between different PAS cells, it is necessary to calibrate the cell effects independently of the sample effects.

Murphy and Aamodt (1977) have developed a photoacoustic thermophone to calibrate PAS cells directly in terms of absolute energy units.

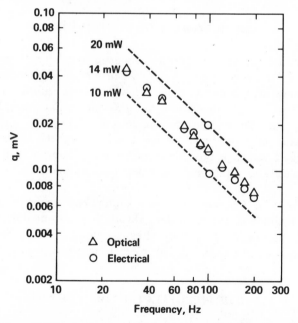

Figure 12.7   Acoustic intensity versus modulation frequency for optical and electrical heating of the metal-black sample in a photothermophone. (Reproduced by permission from Murphy and Aamodt, 1977.)

They mount, as a sample in the PAS cell, an electrically conductive metal black deposited on an insulating substrate. This black serves as a spectrally flat optical absorber. The sample also serves as a uniform source of resistance heating when connected to an external electrical source. When the sample is in the form of a thin film, the cell response under both optical and electrical heating is the same and the frequency-dependent cell effects are identical in both cases. The optical response can then be measured in terms of the electrical input power. Such a device has been used to calibrate microphones and is known as a thermophone (Arnold and Crandall, 1917; Wente, 1922).

Figure 12.7 shows some of the results obtained with the photoacoustic thermophone. Both the thermophone signal arising from the electrical heating of the sample and the photoacoustic signal arising from illumination with an intensity-modulated laser beam exhibit an ω^{-1} dependence as expected. The data indicate that the photoacoustic signal was due to 1.3 mW of heat power emanating from the sample. This is in excellent agreement with the estimated heating that one would expect from a black absorber illuminated by the 1.45-mW laser beam used.

12.6   PIEZOELECTRIC DETECTION

The gas–microphone detector of photoacoustic signals has been found to be very good for many applications, particularly at low modulation frequencies. However, when one is dealing with a large sample or a sample with a small surface/volume ratio, the gas–microphone technique often proves inadequate. Since photoacoustic detection is primarily the detection of the internal heat produced within a sample by the deexcitation of optical energy levels, it is possible to measure this internal heating by the stresses produced in a piezoelectric detector in intimate contact with the sample.

As we show in Chapter 10, stress–strain and acoustic signals are produced in condensed media that absorb intensity-modulated optical radiation. Both the low-frequency stress–strain and the high-frequency acoustic signals can be detected with piezoelectric transducers.

12.7   THE PIEZOELECTRIC TRANSDUCER

Piezoelectricity is the phenomenon related to changes in the physical dimensions of a material with changes in an externally applied electric field. Piezoelectric crystals and ceramics are among the principal detectors and generators of acoustic power. Although piezoelectric crystals such as quartz and Rochelle salt are still widely used for certain applications, most piezoelectric transducers are currently made from polycrystalline ceramics that have been poled or polarized. The most common piezoelectric ceramic is a solution of lead zirconate titanate (PZT), usually modified with other additives.

Piezoelectric materials are ferroelectric in that below a certain temperature, the ferroelectric Curie point, the electric dipoles of individual ions align cooperatively into ferroelectric domains. These domains can be polarized or poled into a particular direction under an external field if the coercivity is not too high. In highly anisotropic single crystals, the polarization vector lies along one of the crystal axis and is very difficult to rotate from that axis. Piezoelectric ceramics exhibit no piezoelectricity when first manufactured since they are formed of a mass of randomly oriented crystallites, each with several ferroelectric domains, and thus the ceramics as a whole are isotropic. They are rendered piezoelectric by a poling treatment, which is the last stage of manufacture and which involves application of a high electric field in a heated oil bath at a temperature not far below the Curie point. Apart from the poling treatment, manufacture of piezoelectric ceramics is similar to that of the more common insulation ceramics, except that closer control is necessary to achieve the desired properties.

All ferroelectric materials are not necessarily piezoelectric. A necessary condition for the occurrence of piezoelectricity is the absence of a center of symmetry. Since piezoelectricity provides a coupling between elastic and dielectric phenomena, piezoelectric properties involve both elastic and dielectric constants.

The electrical condition of an unstressed material placed under the influence of an electric field is defined by two quantities—the field strength E and the dielectric displacement D. Their relationship is

$$D = \chi E \qquad (12.3)$$

where χ is the permittivity of the material.

The elastic condition of the same material at zero electric field strength is defined by the two elastic quantities, the applied stress p and the strain ε. The relationship is

$$p = Y\varepsilon \qquad (12.4)$$

or

$$\varepsilon = sp \qquad (12.5)$$

where s is the elastic compliance of the material ($s = 1/Y$ where Y is Young's modulus).

Piezoelectricity involves the interaction between the electrical and mechanical behavior of the material. In general the piezoelectric condition is described by a 9×9 matrix wherein each column relates to one stress variable (elastic stress component p or electric field E), as independent variable, and each row to a strain variable (elastic strain component ε or electric displacement component D), as dependent variable.

The interaction between electric and elastic properties can be described, in general, by the following linear relations:

$$\varepsilon = s^E p + dE \tag{12.6}$$

$$D = dp + \chi^P E \tag{12.7}$$

Similarly, we can write

$$E = -gp + \frac{1}{\chi^p} D \tag{12.8}$$

$$\varepsilon = s^D p + gD \tag{12.9}$$

In the equations, s^D, s^E, χ^p, d, and g are the principal coefficients. The superscript to the symbols denotes the quantity kept constant under boundary conditions. For example, if by short-circuiting the electrodes, the electric field across a piezoelectric material is kept constant, the superscript E is used. By keeping the electrode circuit open, the dielectric displacement is kept constant, and the superscript D is used. Thus s^D and s^E are the elastic compliances at constant dielectric displacement and at constant electric field, respectively. The term χ^p represents the electrical permittivity at constant elastic stress. The terms d and g are the principal piezoelectric coefficients.

From (12.6)–(12.9), we can define d and g as

$$d = \frac{\varepsilon - s^E p}{E} \tag{12.10}$$

or

$$d = \frac{D - \chi^P E}{p} \tag{12.11}$$

$$g = \frac{D + \chi^P E}{X^p p} \tag{12.12}$$

or

$$g = \frac{\varepsilon - s^D P}{D} \qquad (12.13)$$

For E constant,

$$d = \frac{D}{p} \qquad (12.14)$$

For p constant,

$$d = \frac{\varepsilon}{E} \qquad (12.15)$$

For E constant,

$$g = \frac{D}{X^P p} = \frac{1}{\chi^P} d \qquad (12.16)$$

For ε constant,

$$g = -s^D \frac{p}{D} = \frac{\varepsilon}{D} \qquad (12.17)$$

Thus

$$d = \chi^P g \qquad (12.18)$$

and

$$s^D = (1 - k^2) s^E \qquad (12.19)$$

where

$$k^2 = \frac{d^2}{s^E \chi^P} \qquad (12.20)$$

or

$$\frac{k^2}{1 - k^2} = \frac{g^2 \chi^P}{s^D} \qquad (12.21)$$

In piezoelectric materials, the coefficients depend on the directions of electric field, displacement, stress, and strain. Therefore, subscripts indicating direction are added to the symbols. For piezoelectric ceramics, the

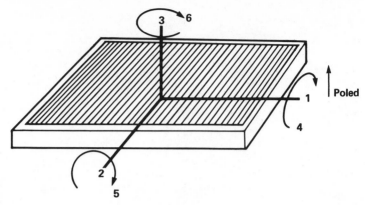

Figure 12.8 A piezoelectric disk showing longitudinal stress directions X, Y, and Z as 1, 2, and 3, and shear stress directions as 4, 5, and 6. The direction of polarization of the disk is Z (3).

z-axis is usually along the direction of positive polarization. If, as shown in Figure 12.8, the directions of X, Y, and Z are represented by 1, 2, and 3, respectively, and the shear about these axis is represented by 4, 5, and 6, respectively, the various related parameters may be written with subscripts referring to these.

For example, χ_{ij}^{p} is the permittivity at constant p when the dielectric displacement (strain) is along the i-axis while the electric field (stress) is along the j-axis. Similarly s_{ij}^{E} is the elastic compliance at constant E when the strain is along the i-axis while the stress is along the j-axis. Note that while χ_{ij} has i and j equal to 1, 2, or 3, s_{ij} can have i and j equal to 4, 5, and 6 as well, since the elastic parameters can have shear–strain and shear–stress as well as uniaxial strain and stress.

For the piezoelectric constants d_{ij} and g_{ij}, the first subscript i refers to the direction of the elastic field or displacement, while the j refers to the direction of the mechanical stress or strain. Note that i can be 1, 2, or 3, while j can be 1, 2, 3, 4, 5, or 6.

For example, $d_{33} = \varepsilon_3 / E_3$ and is the ratio of the strain in the 3 direction to the field applied in the 3-direction, the piezoelectric body being mechanically free ($p = 0$) and not subjected to fields in the 1- and 2-directions ($E_1 = E_2 = 0$). Also $d_{33} = D_3 / p_3$ and thus represents the ratio of the charge per unit area flowing in the 3-direction to the stress applied in the 3-direction when the electrodes are short-circuited ($E = 0$). Again the material is assumed free of all other stresses.

Similarly, $g_{31} = E_3 / p_1$ and is the ratio of the field developed in the 3-direction to the stress applied in the 1-direction when there are no other external stresses and where there are no charges applied ($D=0$). Or $g_{31} = \varepsilon_1/D_3$, which is the ratio of the strain in the 1-direction to the density of the charge applied in the 3-direction when $p=0$.

There are two other common piezoelectric coefficients, the coupling factor k and the frequency constant N_0. The coupling factor k satisfies the relationship

$$k^2 = \frac{\text{stored energy converted}}{\text{stored input energy}} \qquad (12.22)$$

at frequencies below the resonance frequency. This formula holds for both electromechanical and mechanoelectrical energy conversion. A study of k shows that up to 50% of the stored energy can be converted at low frequencies. The value of k is the theoretical maximum, but in practical transducers the conversion is usually lower. Thus k_{ij} is the coupling factor between stored mechanical energy added or converted in the j-direction and the stored electrical energy converted or added in the 3-direction. Thus

$$k_{31} = \frac{d_{31}}{\left(\varepsilon_{33}^p s_{11}^E\right)^{1/2}} \qquad (12.23)$$

The planar coupling factor k_p of a thin disc denotes the coupling between the electric field in the 3-direction (thickness direction) and the simultaneous mechanical actions in the 1- and 2-directions, which results in radial vibration.

$$k_p = k_{31}\left[\frac{2}{(1-\sigma_p)}\right]^{1/2} \qquad (12.24)$$

where σ_p is Poisson's ratio and is the ratio of the compliances in the 1- and 2-directions.

$$\sigma_p^E = -\frac{s_{12}^E}{s_{11}^E} \qquad (12.25)$$

Similarly, a thin disc has a thickness coupling factor k_t, which denotes the coupling between the electric field in the 3-direction (thickness direction) and the mechanical vibration in the 3-direction. Note that k_t is usually smaller than k_{33} because of the constraint imposed by the large lateral dimensions of the disc relative to the thickness. The coupling factor k_t is related to both k_{33} and k_{31} by

$$k_t = \frac{k_{33} - A k_p}{\left[(1 - A^2)(1 - k_p{}^2) \right]^{1/2}} \qquad (12.26)$$

where

$$A = \frac{\sqrt{2}\, s_{13}{}^E}{\left[s_{33}{}^E \left(s_{11}{}^E + s_{12}{}^E \right) \right]^{1/2}} \qquad (12.27)$$

In dynamic systems, the coupling factors are dependent on the stress distribution and are in general less than the static ones because not all the

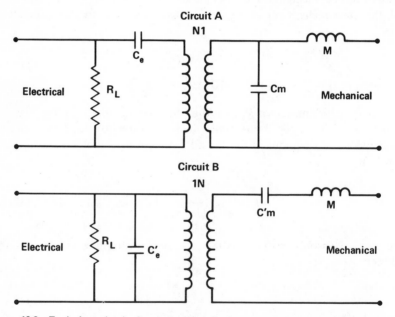

Figure 12.9 Equivalent circuits for a piezoelectric element. Mechanical terminals represent the face or point of mechanical energy to or from the piezoelectric element. The inductance symbol represents an ideal electromechanical transformer—one that transforms voltage to force and vice versa, without loss of energy storage. Transformation ratios, $N:1$ in circuit A and $1:N$ in circuit B, are ratios of voltage input to force output, and also the ratios of velocity input to current output. The capacitance symbols on the electrical side represent electrical capacitances. The capacitance symbols on the mechanical side represent mechanical compliances. (Reproduced by permission from Modern Piezoelectric Ceramics, 1971.)

TABLE 12.1 Electrical Properties of Some Piezoelectric Ceramics.

	Multiplying Factor	PZT-4	PZT-5A	PZT-5H	PZT-8
Coupling Coefficients					
k_{33}	—	0.70	0.71	0.75	0.64
k_{31}	—	0.33	0.34	0.39	0.30
k_{15}	—	0.71	0.69	0.68	0.55
k_p	—	0.58	0.60	0.65	0.51
Piezoelectric Constants					
d_{33}	10^{-10}	285	374	593	225
d_{31}	10^{-10}	−122	−171	−274	−97
d_{15}	10^{-10}	495	584	741	330
g_{33}	10^{-6}	24.9	24.8	19.7	25.4
g_{31}	10^{-6}	−10.6	−11.4	−9.1	−10.9
g_{15}	10^{-6}	38.0	38.2	26.8	28.9
Free Dielectric Constants					
K_1	—	1475	1730	3130	1290
K_3	—	1300	1700	3400	1000
Elastic Constants— Short Circuit					
$1/s_{11}^E = Y_{11}^E$	10^{11}	8.2	6.1	6.1	8.7
$1/s_{33}^E = Y_{33}^E$	10^{11}	6.6	5.3	4.8	7.4
$1/s_{44}^E = Y_{44}^E$	10^{11}	2.6	2.1	2.3	3.1
Density	—	7.6	7.7	7.5	7.6
Mechanical Q	—	500	75	65	1.000
Curie Point	—	325°C	365°C	195°C	300°C

d = resulting strain/applied field in cm/volt
g = resulting field/applied stress in volt-cm/dyne
$1/S = Y$ in dynes/cm^2
Density in g/cm^3

Source: Modern Piezoelectric Ceramics, 1971.

elastic energy is dielectrically coupled. There are exceptions, however, for certain resonant modes of vibration. For example, the fundamental resonant mode of a piezoelectric ceramic in the form of a ring, poled either radially or axially, has no overtones and the static and dynamic coupling factors are identical. In general, however, the dynamic k's are 20–25% lower than the static ones even at resonance. At frequencies far from resonance the coupling factor can be quite low.

The frequency constant N_0 is the product of the resonance frequency and the linear dimension governing the resonance. If the applied electric field is perpendicular to the direction of vibration, then the resonance is the series resonance. If the field is parallel, then it is the parallel resonant frequency, as determined from the equivalent electrical circuit shown in Figure 12.9. Table 12.1 lists some of the electromechanical properties for some typical piezoelectric ceramics.

12.8 PIEZOELECTRIC EXPERIMENTS

A piezoelectric transducer is about 2 orders of magnitude less sensitive than a microphone for a given pressure. However, this is compensated for if one is dealing with large samples, or samples of low surface/volume ratio, or samples in which the heat flow to the surrounding gas is only a small fraction of the total heat generated within the sample. It should be noted that a piezoelectric detection of pressure signals in a gaseous medium would be very inefficient because of the large acoustic impedance mismatch between the gas and the solid transducer.

Acoustic impedance Z_a is given by

$$Z_a = \rho c_0 \qquad (12.28)$$

where ρ is the density and c_0 is the sound velocity. For most solids, Z_a is in the range of 10^6 g/cm^2 sec, while for liquids it is about 10^5 g/cm^2 sec, and for gases Z_a is in the range of 10^2 g/cm^2 sec.

The transmission coefficient for an acoustic wave in a gas going through a gas–solid interface is given by

$$(1-R) = \frac{4Z_g Z_s}{(Z_g + Z_s)^2} \sim 10^{-4} \qquad (12.29)$$

Thus a microphone is a much more sensitive detector of pressure fluctuations in a gas. On the other hand, $(1-R)$ for liquid to solid is $\sim 10^{-1}$ and for solid to solid it is ~1. Thus a piezoelectric transducer is a suitable detector of thermally generated pressure or acoustic fluctuation in condensed materials. Furthermore, piezoelectric detectors can operate at much higher frequencies than microphones.

Several investigators have used piezoelectric detection in photoacoustic experiments. Hordvik and Schlossberg (1977), have used the technique to study laser glasses and windows. Lahmann et al. (1977), Oda et al. (1978), and others have detected photoacoustic signals in liquids with piezoelectric transducers. For the study of solids, one can simply attach the piezoelectric transducer to the sample with wax or a suitable cement as depicted in Figure 12.10. It is imperative that the coupling agent between the sample and the transducer have a good acoustic impedance match with both the sample and the transducer. At very high frequencies ($>$ MHz) it is possible to use a viscous fluid as the coupling agent, but at lower frequencies, where the surface of the sample must be constrained to produce a measurable signal (see Chapter 10), the coupling agent should be a hard solid. In the case of powders one can attach the powder to the transducer with a suitable cement. Care should be taken to prevent any of the incident light from striking the transducer, since this will produce a signal arising from the thermally generated stresses within the transducer and from the changes in the piezoelectric characteristics with temperature. Although the piezoelectric detector is fairly insensitive to airborne noise because of the large

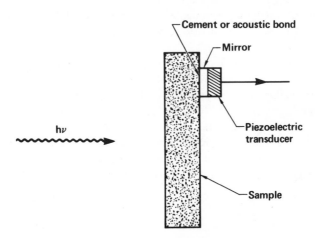

Figure 12.10 A piezoelectric transducer mechanically bonded to a sample with a hard wax or cement.

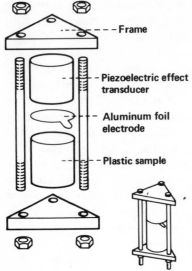

Figure 12.11 A piezoelectric arrangement wherein the transducer is mechanically coupled to the sample by compressional forces. (Reproduced by permission from Callis, 1976.)

acoustic impedance between a gas and solid, the sample and transducer should be well isolated from vibrational noise.

Another method for piezoelectric detection of a photoacoustic signal in solids is shown in Figure 12.11 (Callis, 1976). Here a cylindrical sample, or a sample dispersed in a cylindrical block of plastic, and a cylindrical piezoelectric transducer are clamped rigidly together with an aluminum frame. Photoacoustically induced stress–strain fluctuations in the sample are transmitted to the transducer, with the subsequent generation of a detectable voltage.

For the study of liquids the sample is constrained within a sample chamber, for example, a cylinder, that is in part or in whole comprised of a piezoelectric material. Windows can be bonded at the ends of the cylinder and suitable filling ports drilled through the walls of the tube.

Except for the preamplifier, which generally is a charge or current preamplifier if a piezoelectric ceramic is used as a transducer, the electronic system for these piezoelectric PAS spectrometers is identical to that used in the gas–microphone PAS spectrometers.

REFERENCES

Arnold, H. D., and Crandall, I. B. (1917). *Phys. Rev.* **10**, 22.

Cahen, D., Lerner, E. I., and Auerbach, A. (1978). *Rev. Sci. Instrum.* **49**, 1206.

Callis, J. B. (1976). *J. Res. Natl. Bur. Stand.* **80A**, 413.

Eaton, E. H., and Stuart, J. D. (1978). *Anal. Chem.* **50**, 587 (1978).

Ferrell, W. G., Jr., and Haven, Y. (1977). *J. Appl. Phys.* **48**, 3984.

Gray, R. C., Fishman, V. A., and Bard, A. J. (1977). *Anal. Chem.* **49**, 697.

Hordvik, A., and Schlossberg, H. (1977). *Appl. Opt.* **16**, 101.

Kinsler, L. E., and Frey, A. R. (1962). *Fundamentals of Acoustics*, Chapter 9, Wiley, New York.

Kreuzer, L. B. (1971). *J. Appl. Phys.* **42**, 2934.

Lahmann, W., Ludewig, H. J., and Welling, H. (1977). *Anal. Chem.* **49**, 549.

Modern Piezoelectric Ceramics (1971). PD-9247, Vernitron Piezoelectric Division, Bedford, Ohio.

Murphy, J. C., and Aamodt, L. C. (1977). *Appl. Phys. Lett.* **31**, 728.

Oda, S., Sawada, T., and Kamada, H. (1978). *Anal. Chem.* **50**, 865.

Rosencwaig, A. (1977). *Rev. Sci. Instrum.* **48**, 1133.

Sessler, G. M., and West, J. E. (1966). *J. Acoust. Soc. Am.* **40**, 1433.

Wente, E. C. (1922). *Phys. Rev.* **19**, 333.

PHOTOACOUSTIC EXPERIMENTS
WITH LIQUIDS

13.1 INTRODUCTION

There has been considerable interest in the possibility of using the photo-acoustic technique for investigating optical absorption processes in liquids. This interest is stimulated by two problems that cannot be effectively managed by conventional spectrophotometry: (*1*) the accurate measurement of a weakly absorbing solution; and (*2*) the analysis of highly light-scattering liquid systems such as suspensions. It is at present quite difficult to measure absorption coefficients in solutions where the absorption coefficient is much less than 10^{-3} cm^{-1}. It is also most difficult to obtain reliable absorption spectra when dealing with highly light-scattering liquid suspensions. The problem of dealing with low absorption solutions has been partially resolved by the use of fluorescence techniques, and in fact it has been shown that laser fluorescence detection combined with chromatographic separation can yield a detection limit of $\sim 7.5 \times 10^8$ molecules rhodamine 6G/cm^3 or an effective absorption coefficient of $\sim 10^{-7}$ cm^{-1} (Fairbank et al., 1975). However, in trace analysis, the analytical procedure often involves the use of a nonfluorescent, highly absorbing indicator for the substance to be measured. Here fluorimetry is not possible, and conventional spectrophotometric analysis permits a sensitivity of only 10^{-3} cm^{-1}. In the meantime, the problem of highly light-scattering solutions is still not well resolved.

13.2 GAS-MICROPHONE METHOD

There are two ways by which a photoacoustic signal can be obtained from an illuminated solution. In the first method, the solution is itself a sample, much as a solid material, in a conventional photoacoustic cell. The periodic heat generated within the solution by optical absorption processes diffuses to the liquid–gas interface, perturbs the gas within a boundary layer or so of the interface, and creates an acoustic disturbance detected by a conventional gas mcrophone. The theory for this effect is identical to that developed in Chapter 9, with the appropriate liquid parameters inserted where the sample parameters are needed. Table 9.1 gives, for example, some of these parameters, such as the thermal diffusivity and heat conductivity, for water and other liquids.

Several investigators have studied optical absorption in solutions by this technique (McClelland and Kniseley, 1976; Wetsel and McDonald, 1977; Gray et al., 1977). In general, these investigators have found that photo-acoustic signals are readily detectable from solutions with optical absorption coefficients as low as 0.1 cm^{-1}.

13.3 PIEZOELECTRIC METHOD

To achieve a higher sensitivity for photoacoustic measurements in liquids, some researchers have investigated the possibility of using a piezoelectric transducer that couples directly to the thermoelastic waves produced in the liquid. The gas microphones that are condenser-type microphones are not suitable for measurements of pressure changes in liquids since they are sensitive primarily to the extent of motion of the microphone membrane. Since the amplitude of motion in an elastic medium is given by

$$\Delta x = \frac{\Delta p}{2\pi\rho v} \tag{13.1}$$

where Δp is the pressure fluctuation, ρ is the density of the fluid, and v is the frequency of the pressure fluctuation, it is readily seen that Δx for liquids is $\sim 10^{-3}$ smaller than for gases undergoing the same pressure fluctuation.

As we see in Chapter 10, thermoelastic signals generated by the photo-acoustic effect can be readily detected in liquids with piezoelectric trans-ducers, provided that the liquid sample is constrained within a pressure chamber.

There has, as yet, been very little work reported on the photoacoustic effect in liquids where a liquid-coupling piezoelectric transducer has been used. Nevertheless, not only is the technique feasible, but as Lahmann et al. (1977) have shown, it is highly practical, especially in the case of a weakly absorbing solution.

In Figure 13.1, we show a schematic of their experimental arrangement. In a photoacoustic experiment on a weakly absorbing species in a solvent, there is often a strong background signal from absorption in the solvent itself. To overcome this problem, Lahmann et al. separated the output of an argon ion laser operating in the "all lines mode" by the prism Pl into its

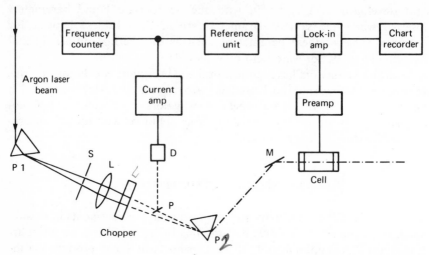

Figure 13.1 Schematic diagram for the experimental arrangement used in the liquid photo-acoustic experiment described in the text. (Reproduced by permission from Lahmann et al., 1977.)

spectral components. A perforated plate S transmits the two most powerful beams at 488 and 514 nm and screens off all the others. The two beams are reunited with the lens L and the prism $P2$, and the recombined beam is directed into the sample cell. A chopping wheel, situated after the lens L, interrupts the two beams such that the laser beam striking the sample cell has a slightly varying intensity at twice the chopping frequency but with a switching wavelength. In their work Lahmann et al. operated the laser at 700 mW power and it was chopped at 700 Hz. This dual-wavelength operation enables a substantial reduction of the background signal, which is considerable even for solvents having absorption coefficients as low as 10^{-3} or 10^{-4} cm^{-1}. By working with solvents that have a weak and fairly flat absorption profile in the wavelength region used, one can reduce the background signal down to a residual absorption coefficient in the neighborhood of 10^{-5} cm^{-1}. This then permits the detection of a solute concentration so low that the net absorption coefficient differential between the wavelength regions used is only 10^{-5} cm^{-1}. In the experiment of Lahmann et al., a detection limit of about 0.08 ng/cm^3 of β-carotene (2.2×10^{-5} cm^{-1}) was found. They also obtained a detection limit of 15 ng/cm^3 for selenium in chloroform corresponding to an absorption coefficient of 3.5×10^{-5} cm^{-1}.

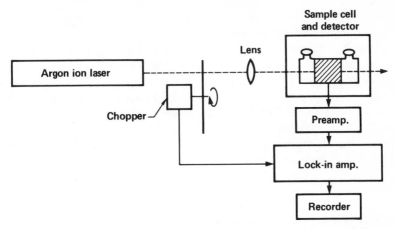

Figure 13.2 Block diagram of a laser-driven liquid photoacoustic spectrometer. (Reproduced by permission from Oda et al., 1978.)

A simpler and more conventional liquid PAS setup has been used by Oda et al. (1978) to determine the presence of trace amounts of the metal cadmium in a chloroform solution. Their apparatus is shown in Figure 13.2.

Using a 500-mW argon-ion laser, Oda et al. found that the PAS signal was linear in the concentration of Cd in chloroform from 0.05 to 50 ng/ml Cd. In comparison to colorimetric and conventional atomic absorption spectrometry, the linear concentration range offered by the PAS technique was at least 1 order of magnitude greater. Moreover, the range could probably be further increased with an increase in the laser power.

The detection limit for the system used by Oda et al. was estimated to be 0.02 ng/ml Cd (14 ppt), based on a limiting photoacoustic signal/noise ratio of 2:1. This value is about 2 orders of magnitude lower than that for colorimetric analysis (3 ng/ml) (Saltzman, 1953) and for conventional phase atomic absorption measurements (10 ng/ml) (Slavin, 1968).

The piezoelectric photoacoustic technique is clearly applicable to the study of weak absorptions in liquids in the UV, visible, or IR with appropriate light sources. However, some further development is needed in eliminating the absorption of the unwanted signals from scattered light by the walls of the cell and in treating the inner surfaces of the piezoelectric ceramic tubes, so that compounds dissolved in concentrated acidic and basic media can be measured. One possibility might be to coat the inner surfaces of these piezoelectric tubes with highly reflective gold overlayed with a transparent inert plastic.

When the incident light intensity is very high, two-photon absorption can be measured as well as ordinary single-photon absorption. This has been demonstrated by Bonch-Bruevich et al. (1977). In their experiment a pulsed-dye laser operating in the nanosecond region was used to illuminate a polymethine dye solution and an ethanol solution of anthracene contained in a piezoelectric cell.

When the absorption is by single-photon processes only, the amplitude of the piezoelectric signal per laser pulse is given by

$$q(\nu) = AI(\nu)\tau\eta(\nu)(1 - e^{-\beta(\nu)l}) \qquad (13.2)$$

where A is a constant determined by calibrations, $I(\nu)$ is the intensity of the laser pulse, τ is the pulse length, ν is the optical frequency, η is the quantum yield of nonradiative transitions, β is the optical absorption coefficient, l is the optical path length in the cell, and $E(\nu)$ is the incident energy. The constant A is a function of the cell and piezoelectric transducer properties and the ultrasonic attenuation of the solution. To obtain a value for A, one must make an opaque solution having the same ultrasonic

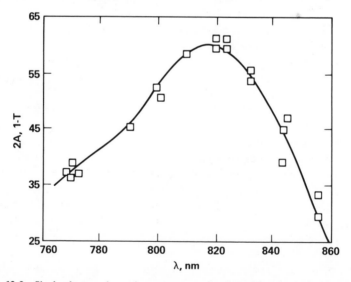

Figure 13.3 Single-photon absorption spectrum of polymethine dye solution in ethanol. (___) Absorption spectrum obtained with a spectrophotometer; squares (□) absorption spectrum measured with photoacoustic spectrometer. (Reproduced by permission from Bonch-Bruevich et al., 1977.)

properties as the experimental solution and accurately measure $q(\nu)$ and the quantum yield $\eta(\nu)$ as a function of $I(\nu)$. In Figure 13.3 the single-photon photoacoustic spectrum of polymethine dye solution in ethanol is compared with a conventional absorption spectrum obtained on a spectrophotometer.

If the laser intensity is sufficiently high, multiphoton absorption can occur, and the piezoelectric signal is then given by

$$q(\nu) = \sum_{m=1}^{n} A\left[I(\nu) \right]^{m} \eta(\nu_m)\rho(\nu)Nl \qquad (13.3)$$

where $\eta(\nu_m)$ represents the quantum efficiency for nonradiative transitions of the ν_m level, $\rho(\nu)$ is the absorption cross section, and N is the density of absorbing molecules. In (13.3), we assume that the solution is optically thin and that

$$(1 - e^{-\beta l}) \simeq \beta l = \rho(\nu)Nl \qquad (13.4)$$

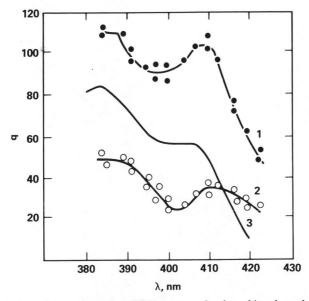

Figure 13.4 Two-photon absorption (TPA) spectra of polymethine dye solution. (*1*) TPA spectrum, measured by luminescence intensity, (*2*) TPA spectrum, measured by photoacoustics; (*3*) single-photon absorption spectrum. The x-axis gives wavelengths corresponding to double the radiation frequency of the tunable laser. (Reproduced by permission from Bonch-Bruevich et al., 1977.)

For the intensities available with most dye lasers, the two-photon absorption ($m=2$) predominates at high intensities. Bonch-Bruevich et al. obtained the two-photon absorption of the polymethine dye, with the final states between 380 and 430 nm, by operating at high intensities in the 760–860 nm range of the laser. This is shown in Figure 13.4. Also shown is the two-photon absorption for this region as determined through measurements of the luminescence intensity (curve 2) and the conventional single-photon absorption in this wavelength region (curve 3). There is good agreement between the photoacoustic and luminescence measurements of two-photon absorption. It should be noted that the considerable similarity of the single and two-photon absorption spectra for the polymethine dye solution apparently indicates that strict selection rules are lacking in this complex molecule with its low symmetry, thus allowing single-photon and two-photon transitions between the same states.

The dependence of the piezoelectric signal and that of the luminescence intensity for an anthracene solution in ethanol (absorption band in the

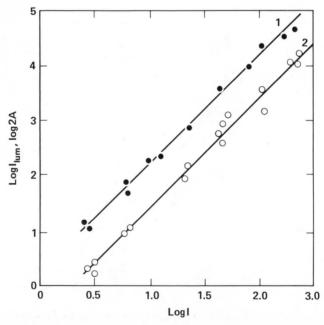

Figure 13.5 Dependence of the signal amplitude of the photoacoustic piezoelectric transducer (curve 2) and of the luminescence intensity (curve 1) of an anthracene solution in ethanol on the intensity of the exciting radiation. (Reproduced by permission from Bonch-Bruevich et al., 1977.)

350–380 nm range) as a function of the intensity of the laser radiation at 694.3 nm is plotted in Figure 13.5. As can be seen from this figure, both amplitudes are quadratic functions of the excitation intensity, as would be expected for a two-photon absorption.

The photoacoustic method allows the measurement of both single-photon and multiphoton absorptions. Conventional absorption measurements of multiphoton processes are difficult because of the weak absorptions due to these processes. Although one can use luminescence measurements to investigate multiphoton events in fluorescent materials, photoacoustic spectroscopy can be used for nonfluorescent materials as well.

REFERENCES

Bonch-Bruevich, A. M., Razumova, T. K., and Starobogatov, I. O. (1977). *Opt. Spectrosc.* **42**, 45.

Fairbank, W. M., Hansch, T. W., and Schalow, A. L. (1975). *J. Opt. Soc. Am.* **65**, 199.

Gray, R. C., Fishman, V. A., and Bard, A. J. (1977). *Anal. Chem.* **49**, 697.

Lahmann, W., Ludewig, H. J., and Welling, H. (1977). *Anal. Chem.* **49**, 549.

McClelland, J. F., and Kniseley, R. N. (1976). *Appl. Phys. Lett.* **28**, 467.

Oda, S., Sawada, T., and Kamada, H. (1978). *Anal. Chem.* **50**, 865.

Saltzman, B. E. (1953). *Anal. Chem.* **25**, 493.

Slavin, W. (1968). *Atomic Absorption Spectroscopy*, Interscience, New York.

Wetsel, G. C., Jr., and McDonald, A. F. (1977). *Appl. Phys. Lett.*, **30**, 252.

SPECTROSCOPIC STUDIES

14.1 INTRODUCTION

Even though photoacoustic spectroscopy of condensed matter has been actively pursued for only a few years, it has already demonstrated its power as a spectroscopic tool in physics and chemistry. In this chapter we explore some of the spectroscopic applications of photoacoustics in the physical and chemical sciences.

14.2 INORGANIC INSULATORS

When the surface of a solid material is not highly reflective, photoacoustic spectroscopy can provide optical data about the bulk material itself. The PAS technique can thus be used to study insulator, semiconductor, and even metallic systems that cannot be studied readily by conventional absorption or reflection techniques, for example, substances that are in the form of powders, or those that are amorphous, or those that for some reason are difficult to prepare for reflection studies.

In the case of insulators, photoacoustic spectra give direct information about the optical absorption bands in the material. This is shown in Figure 14.1. Spectrum a shows the normalized PAS spectrum of some Cr_2O_3 powder in the region of 200–1000 nm (Rosencwaig, 1973). Spectrum b is an optical absorption spectrum obtained by McClure (1963) on a Cr_2O_3 crystal 4 microns thick. Spectrum c is a diffuse reflectance spectrum of Cr_2O_3 powder obtained by Tandon and Gupta (1970). The two crystal field bands of the Cr^{3+} ion at 600 and 460 nm are almost as clearly resolved in the photoacoustic spectrum of the Cr_2O_3 powder as they are in the absorption spectrum of the Cr_2O_3 crystal, and they are much better resolved in the PAS spectrum than in the diffuse reflectance spectrum.

Figure 14.2 shows the photoacoustic spectrum of another inorganic insulator CoF_2 (Rosencwaig, 1978). Again this material is in the form of a highly light-scattering powder. We note that both the crystal field bands of the Co^{2+} and the charge-transfer band are clearly visible.

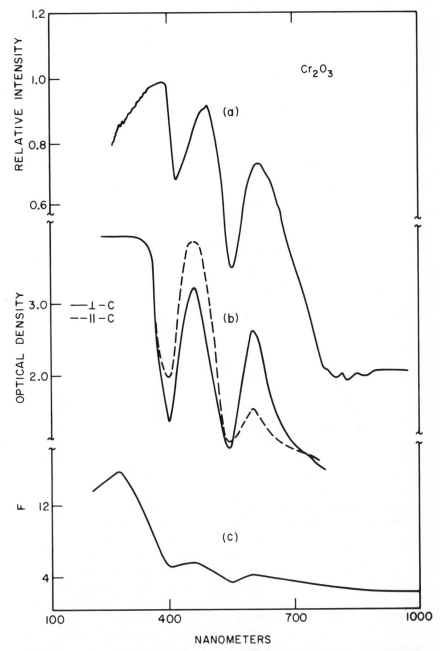

Figure 14.1 (a) The normalized photoacoustic spectrum of Cr_2O_3 powder at 300°K. (b) The optical transmission spectrum of a 4.4-μm-thick crystal of Cr_2O_3 at 300°K. (c) The diffuse reflectance spectrum of Cr_2O_3 powder at 300°K. (Reproduced by permission from Rosencwaig, 1973.)

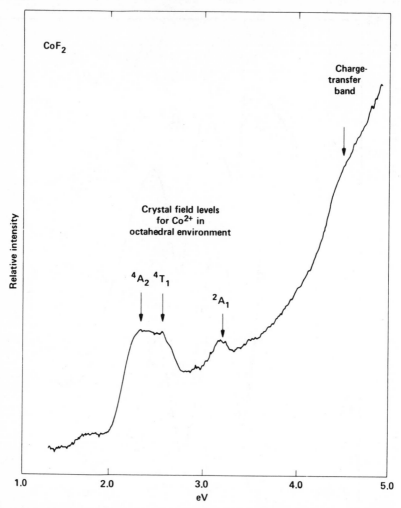

Figure 14.2 The photoacoustic spectrum of CoF_2 powder at 300°K showing both crystal-field and charge-transfer bands. (Reproduced by permission from Rosencwaig, 1978.)

Photoacoustic spectroscopy has also been employed to characterize various thalium iodide (TlI) powders used in thin-film antireflection coatings for high-power CO_2 laser windows (Fernelius, 1978). Wavelength-dependence studies were made around the absorption edge at 440 nm. Two general types of features were observed. The spectrum of what is considered as good coating material displayed a sharp band edge as shown in Figure 14.3. The spectrum of the lower-grade material exhibited a more

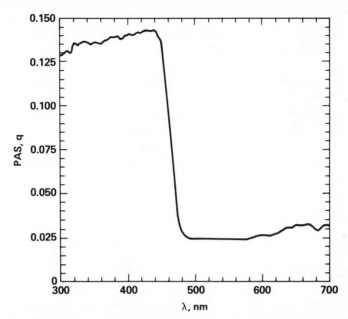

Figure 14.3 PAS spectrum of high-quality TlI powder material. (Reproduced by permission from Fernelius, 1978.)

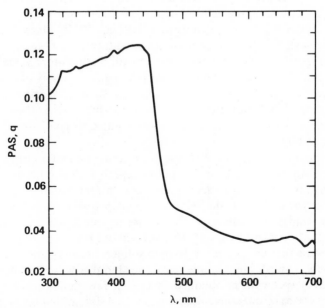

Figure 14.4 PAS spectrum of low-quality TlI powder material. (Reproduced by permission from Fernelius, 1978.)

gradual slope of the band edge as shown in Figure 14.4. This change in spectral characteristics could be correlated to the presence of absorbed I_2 within the solid, which evolved naturally upon exposure to atmosphere. Thus it was possible with PAS not only to identify materials that would result in poor optical coatings, but also to account for the underlying cause of this problem.

Other studies on powders include an investigation of uranium tetrafluoride, one of the products produced in a uranium plasma. Uranium tetrafluoride cannot be readily identified nondestructively by conventional means, but has proved amenable to spectroscopic analysis with photoacoustics (Eaton et al., 1977).

14.3   INORGANIC AND ORGANIC SEMICONDUCTORS

In the case of semiconductors, both direct and indirect band transitions can be observed as shown in Figures 14.5 and 14.6. In Figure 14.5, results are shown for three direct-band semiconductors, all in powder form (Rosencwaig, 1975a; 1975b). The band edge as measured by the position of the knee in the PAS spectra agrees very well with the values recorded in the literature and given in parentheses (Pankove, 1971). Figure 14.6 shows the PAS spectrum of the indirect-band semiconductor GaP, also in the form of a coarse powder (Rosencwaig, 1975a; 1975b). Here again, the band edge value for the forbidden direct-band transition obtained from the PAS measurement agrees remarkably well with the literature value (Nelson et al., 1964). The close agreement is surprising, since one would expect a lower value for the knee as a result of the opacity effects, arising from the strong indirect-band transitions in this semiconductor. It is these indirect-band transitions that give rise to the more gradual slope of the absorption edge seen in Figure 14.6.

Several points concerning PAS spectra of semiconductors can be made. First, the PAS technique gives the correct spectrum for both direct (CdS) and indirect (GaP) bandgap semiconductors. In fact, Somoano (1978), has shown that for GaP both the indirect bandgap transition at 562 nm and the direct bandgap transition at 470 nm are revealed. Second, the samples can consist of powders taken off the shelf without further purification or of single crystals that are gound up to reduce the reflectivity and increase the surface area. Third, the sample mass need only be a few milligrams. Fourth, each spectrum is obtained in only a few minutes. These points emphasize the ease and convenience with which data of importance for the

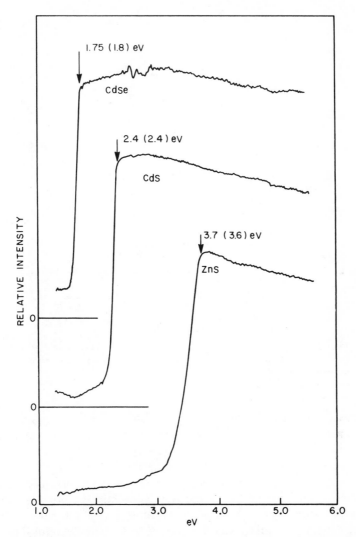

Figure 14.5 Photoacoustic spectra of three direct-band semiconductors in powder form at 300°K. The band gaps derived from these spectra are shown and compared to the values derived from specular reflectance measurements (in parentheses). (Reproduced by permission from Rosencwaig, 1975a; 1975b.)

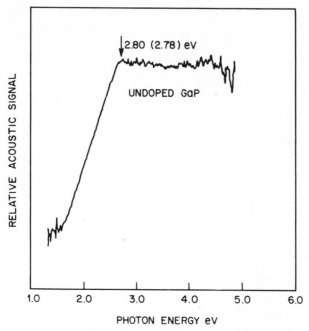

Figure 14.6 Photoacoustic spectrum of the indirect-band semiconductor GaP at 300°K. The shallower slope is a result of the indirect-band transitions. (Reproduced by permission from Rosencwaig, 1975a.)

optical and electrical properties of semiconductors may be obtained without the requirements of high purification or high vacuum techniques. In addition, one can differentiate between different crystalline phases of the same materials, such as the rutile and anatase phases of TiO_2 (Adams et al., 1976). The above semiconductors are of interest for use as electrodes in photoelectrochemical cells to convert solar energy into electrical and chemical energy. Thus PAS provides a convenient tool for the physicist or chemist to quickly screen, characterize, and correlate the optical properties of numerous semiconductors for use in many different systems.

Organic and organometallic semiconductors may also be investigated using PAS. These materials are of interest in the field of quasi-one-dimensional (1-d) conductors. The compounds consist of weakly coupled chains of molecular units that may exhibit very high electrical conductivity parallel to the chains. The compounds may be metallic if the atoms or molecules are uniformly spaced along the chain. However, quasi-one-dimensional metallic chains are fundamentally unstable with respect to certain lattice distortions (Peierls' instability). Thus in some compounds the chains distort in such a manner that the atoms or molecules form dimers, trimers, tetramers, and so on. This distortion, or transition, can occur at temperatures above or below room temperature. The compounds with distorted chains are semiconductors since the nonuniform spacing of the molecules along the chain opens up a gap at the Fermi level in the electronic energy spectrum. Thus the electrical and magnetic properties are crucially dependent on the chain structure. The compounds are usually darkly colored powders obtained most readily as precipitates from chemical reactions. Single crystals, which are difficult to obtain, are often in the form of very small needles (a few millimeters long), making conventional optical studies, even on the crystals, very difficult.

As example of a study of 1-d compounds is given in Figures 14.7 and 14.8. This is a photoacoustic study of a series of iridium carbonyl compounds that contain either the semiconducting linear chains of square-planar cis $-[Ir(CO)_2X_2]$ or the nonchain complex $[Ir(CO)_2X_2]^-$, where X is Cl or Br (Rosencwaig et al., 1976). The photoacoustic spectra seen in Figure 14.7 show three absorption bands below 650 nm at 2.3, 2.9, and 3.4 eV. These bands are assigned as metal-to-ligand charge-transfer transitions from the $a(yz)$ and $b(xz)$ metal orbitals to the predominantly ligand CO $b(\pi^*, 6p_z)$ orbital. At wavelengths longer than 650 nm, the spectrum of the semiconducting materials rise strongly toward the infrared region. This rise is the high-energy end of a broad absorption band extending from 0.1 to 2 eV, as has been shown by conventional infrared transmission spectroscopy. This band has been assigned as a transition from the $5d_{z^2}$ band to the $b(\pi^*, 6p_z)$. As can be seen in Figure 14.7, the nonchain complex (e) does not have this infrared band. In addition, all the observed linear-chain trasitions seen in Figure 14.7 are considerably red-shifted with respect to the corresponding transitions in the nonchain $[Ir(CO)_2Cl_2]^-$. This red shift is attributed to interactions along the chain that raise the energy of $a(yz)$, $b(xz)$, and $a(z^2)$ while lowering the energy of $b(\pi^*, 6p_z)$.

The results shown in Figure 14.8 indicate that the width and energy maximum of the infrared transition depend on the chain length, becoming broader and shifting to higher energy as the chains are shortened by the process of crushing the samples. Observation of this effect gives us a valuable means for determining chain lengths, and thus of correlating chain lengths with observed electronic transport properties.

Another example of the use of PAS in the study of organometallic compounds is given in Figure 14.9 (Somoano, 1978), which shows the PAS spectrum of an interesting rhodium-"bridged" dimer compound (Lewis et al., 1976), in which the rhodium atoms are physically bound into a dimeric

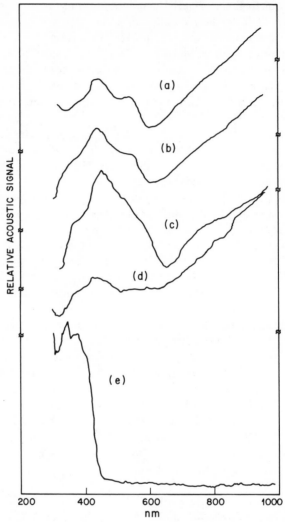

Figure 14.7 Photoacoustic spectra of "as made" samples of (a) $K_{0.98}Ir(CO)_2Cl_{2.42}0.2$ CH_3COCH_3, (b) $K_{0.60}Ir(CO)_2CL_2 \cdot 0.5H_2O$, (c) $(TTF)_{0.61}Ir(CO)_2Cl_2$, (d) $Cs_{0.61}Ir(CO)_2Br_2$; (e) $(C_6H_5)_4As[Ir(CO)_2Cl_2]$. (Reproduced by permission from Rosencwaig et al., 1976.)

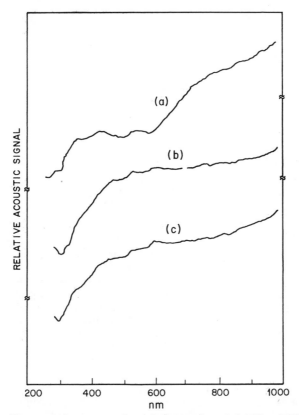

Figure 14.8 Photoacoustic spectra of crushed samples of (a) $K_{0.98}Ir(CO)_2Cl_{2.42}$, (b) $(TTF)_{0.61}Ir(CO)_2Cl_2$, (c) $Cs_{0.61}Ir(CO)_2Br_2$. These spectra show the increased infrared absorption due to smaller chain lengths. (Reproduced by permission from Rosencwaig et al., 1976.)

structure by bridging ligands. The bridged dimers are not part of a chain structure in this compound. The interesting feature is the presence of the strong absorption band at 575 nm, which is due to the dimeric structure. A similar band is found in the solution spectrum. Figures 14.9b and 14.9c show PAS spectra of two rhodium chain compounds. The rhodium atoms are square-planar coordinated by isocyanate ligands, (R-CN) and stack to form rhodium chains in contrast to the rhodium-"bridged" dimer. Nevertheless, the PAS spectrum of rhodium (phenylisocyanate)$_4$ tetraphenyl-borate (Figure 14.9b) reveals the same strong rhodium–rhodium dimer band at 575 nm, indicating a dimeric chain structure. The room temperature electrical conductivity of this material is quite low ($\sigma_{RT} \sim$ $10^{-10}\,\Omega^{-1}cm^{-1}$), as would be expected from the nonuniform (dimeric)

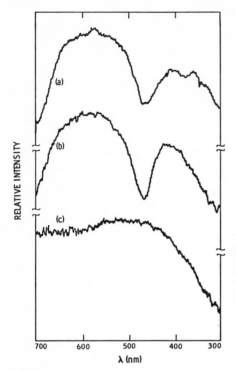

RELATIVE INTENSITY

(a)

(b)

(c)

700   600   500   400   300

λ (nm)

Figure 14.9 Photoacoustic spectra of several rhodium organometallic compounds. (a) Rhodium-"bridged" dimer. (b) Rhodium (phenylisocyanate)$_4$ tetraphenylborate, [Rh(ϕ-CN)$_4$]Bϕ_4. The absorption band at 575 nm in a and b is associated with the formation of dimers. (c) Rhodium (vinylisocyanate)$_4$ perchlorate, [Rh(VI-CN)$_4$]ClO$_4$. (Reproduced by permission from Somoano, 1978.)

spacing along the chain. In contrast, the PAS spectrum of rhodium (vinylisocyanate)$_4$ perchlorate (Figure 14.9c) does not reveal any sign of the dimer, suggesting that this material contains chains of uniformly spaced rhodium atoms. Indeed, the electrical conductivity of this compound ($\sigma_{RT} \sim 2\,\Omega^{-1} - cm^{-1}$) is the highest known for any rhodium chain complex (Williams et al., 1977). All the structural information deduced from the PAS spectra of Figure 14.9 has been fully confirmed by single-crystal X-ray diffraction studies (Williams et al., 1977).

In summary, PAS provides a very simple technique for gaining insight into the structural and electrical properties of quasi-one-dimensional conductors, eliminating the necessity to grow single crystals or perform X-ray measurements. In this way, many new compounds may be screened without the expenditure of excessive time and funds.

In addition to the above experiments, photoacoustic spectroscopy can also be used at cryogenic temperatures and at higher resolution (e.g., with the use of lasers) to study excitonic and other fine structure in crystalline, powder, or amorphous semiconductors, and thus to investigate the effects of impurities, dopants, and electromagnetic fields.

14.4 PHOTOACOUSTIC SATURATION

In the work on semiconductors and certain organic dyes the problem of photoacoustic saturation becomes acute. For here the absorption coefficient β is 10^4 cm^{-1} or greater, and even finely ground particles are optically opaque, although thermally thin. According to the RG theory, such compounds will also be photoacoustically opaque (or saturated) unless the thermal diffusion length μ is less than the optical absorption length β^{-1}. Since it would require modulation frequencies in excess of 10^4 Hz to bring these particles out of saturation ($\mu < 1/\beta$), such samples appear to be inherently opaque, even from a photoacoustic point of view. However, recent work by Lin and Dudek (1979) shows that proper sample preparative techniques can bring even these materials out of saturation. Lin and Dudek develop three separate methods for working with highly opaque systems.

1. A thin layer of a sample is deposited onto a quart plate. The sample deposition can be made either by vacuum deposition or by smearing with a suitable applicator.

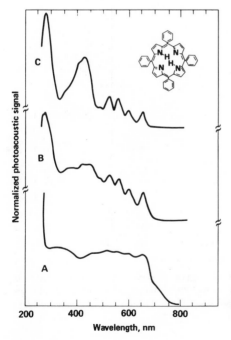

Figure 14.10 Photoacoustic spectra of tetraphenylporphin (TPP). (*A*) Excess powder, (*B*) 2600-A layer on quartz, (*C*) 520-A layer on quartz. (Reproduced by permission from Lin and Dudek, 1979.)

2. The sample can be molecularly dispersed over the surface of a nonabsorbing medium such as barium sulfate, magnesium oxide, neutral alumina, and silica gel. The sample is first dissolved in a suitable solvent such as chloroform. The solution is then thoroughly mixed with an appropriate amount of the nonabsorbing substrate material to from a slurry, and then the solvent is evaporated with a rotary evaporator.

3. A mechanical dispersion technique can be used wherein the sample is coground with a nonabsorbing substrate in a vibrating mill.

Figure 14.10 shows PAS spectra of tetraphenylporphin (TPP) taken by Lin and Dudek. Spectrum A is of TPP powder, spectrum B is of a layer of TPP $\sim$2600 Å thick on a quartz plate, and spectrum C is of a layer of TPP $\sim$520 Å thick deposited on a quartz plate. Saturation is clearly evident in spectrum A of the powder, slightly evident in the 2600 Å thick layer, and apparently not present in the 520 Å thick layer, since all the absorption bands are clearly visible and in the correct ratio.

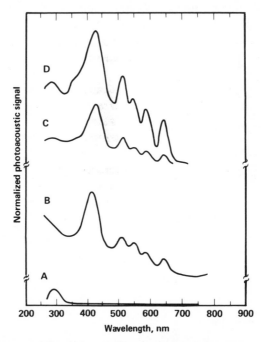

Figure 14.11 Photoacoustic spectra of TPP chemically dispersed by slurry method onto $BaSO_4$ particles. (A) bare $BaSO_4$, (B) 0.001 wt % TPP on $BaSO_4$, (C) 0.01 wt % TPP on $BaSO_4$, (D) 0.1 wt % TPP on $BaSO_4$. (Reproduced by permission from Lin and Dudek, 1979.)

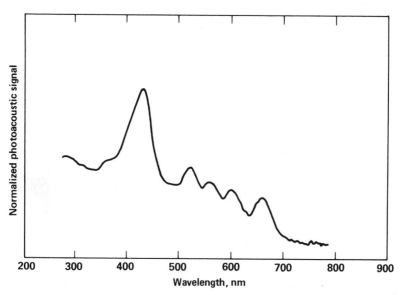

Figure 14.12 Photoacoustic spectrum of 0.35 wt % TPP mechanically dispersed in $BaSO_4$ by cogrinding. (Reproduced by permission from Lin and Dudek, 1979.)

Figure 14.11 gives the results of Lin and Dudek for TPP chemically dispersed on $BaSO_4$. Curve A is of bare $BaSO_4$, showing only a weak absorption in the UV. Curves B, C, and D are of 0.001, 0.01, and 0.1% TPP (by weight in the slurry) dispersed in the $BaSO_4$. All these spectra show the characteristic absorption peaks of TPP without serious saturation effects. The fact that TPP can be detected at a concentration as low as 0.001% indicates that PAS is a very useful tool for characterizing minute quantities of highly absorbing substances. The light scattering from the barium sulfate particles does not interfere with this measurement. This is in sharp contrast with conventional optical measurements, where opacity and light-scattering of the dispersing medium present serious analytical problems.

Finally, Figure 14.12 shows the PAS spectrum of 0.35% TPP mechanically dispersed by cogrinding in $BaSO_4$. Again, a good spectrum free of saturation is obtained.

14.5 METALS

PAS may be used to study metals if the reflectivity is first diminished by grinding or through the use of powders. Because of the very large absorp-

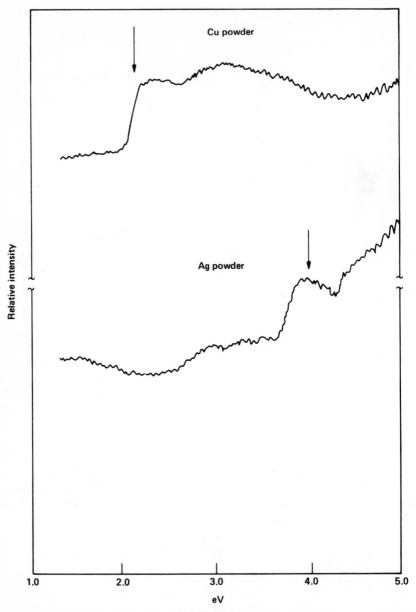

Figure 14.13 Photoacoustic spectra of copper and silver powders. The arrows indicate positions of known structure in the reflection loss spectra of these metals. (Reproduced by permission from Rosencwaig, 1975a; 1975b.)

tion coefficients of metals, the absorption spectrum cannot be obtained by PAS. However, the presence of the metal (e.g., on a surface of a nonmetal) may be detected by its loss of reflectance, which may result from interband transitions, a plasma resonance, and so on. Figure 14.13 shows the PAS spectra of copper and silver powders (Rosencwaig, 1975a; 1975b). The sharp structure in the PAS spectra indicated by the arrows occurs at the point of reflection loss and can serve as a signature for the presence of a particular metal. The structure in these PAS spectra also emphasizes the care one must take in the choice of materials for the use in the construction of a PAS cell.

As in the case of semiconductors, a major advantage of PAS over conventional reflection spectroscopy is that highly reflecting surfaces are not needed and in fact are undesirable for bulk studies. Furthermore, the ultraclean surfaces and thus the elaborate high-vacuum equipment so necessary for conventional reflection studies are not needed for photoacoustic spectroscopy.

14.6 LIQUID CRYSTALS

Organic liquid crystals are another class of materials of considerable interest to both physicists and chemists (DeGennes, 1974). Because the solid, smectic, and nematic states of liquid crystals are highly light scattering, optical spectroscopy on these states cannot be readily performed. Nevertheless optical data on these states may be very useful in providing information about the intermolecular interactions that play so important a role in determining the unusual properties of these materials.

In Figure 14.14a we show data from a photoacoustic study (Rosencwaig, 1975b) of a class of liquid crystals. Since all members of this class are composed of molecules with essentially the same benzylidene aniline center, but with different heads and tails, the general features of the UV PAS spectra are quite similar. However, there is a clear shift in the position of the primary absorption edge as we proceed from MBBA to CBOOA. This shift is probably due to the different heads and tails and thus is probably a measure of the different intermolecular interactions, as is seen in Figure 14.14b, where we have plotted the temperatures at which these materials crystallize and at which they go into their isotropic liquid state along the y-axis, and the value $1/\varepsilon$, where ε is the position of the primary absorption edge, along the x-axis. Thus photoacoustic data on these liquid crystals can be used to obtain information about the intermolecular interactions in these crystals and therefore enhance our understanding of the unusual properties of these compounds.

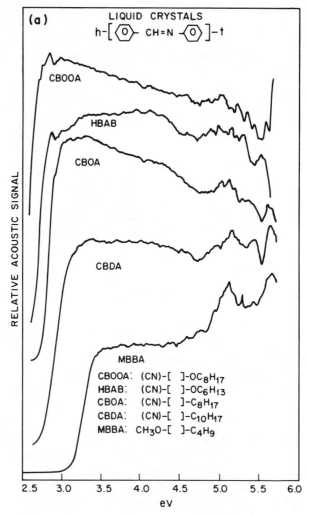

Figure 14.14a A study of five liquid crystals with a benzylidine aniline center section. Their UV PAS spectra are shown in *a*. (Reproduced by permission from Rosencwaig, 1975a.)

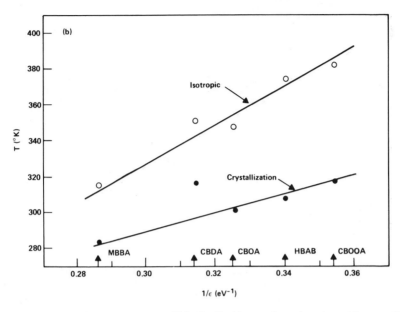

Figure 14.14b The temperatures at which the liquid crystals undergo transitions to their isotropic form, and to their crystal form, are plotted as functions of the reciprocal absorption edge energies. (Reproduced by permission from Rosencwaig, 1975a.)

14.7   INFRARED STUDIES

The experiments reported so far in this chapter have been performed in the UV and visible regions of the optical spectrum. Nordal and Kanstad (1977) have performed studies on ammonium sulfate and glucose powders and on their aqueous solutions in the midinfrared using a CO_2 laser. Figure 14.15 shows their results on $(NH_4)_2SO_4$. Although water also absorbs strongly in the CO_2 wavelength region, Nordal and Kanstad were able to detect a concentration of 10^{-4} $(NH_4)_2SO_4$ by using a differential method whereby the laser alternates between two of the CO_2 laser lines where the absorption of the ammonium sulfate differs significantly, while that for water differs only slightly.

14.8   SPECTROSCOPIC EXPERIMENTS WITH PIEZOELECTRIC DETECTION

As we state earlier, photoacoustic spectroscopy with gas-microphone detection methods is quite effective with powders and solids having large

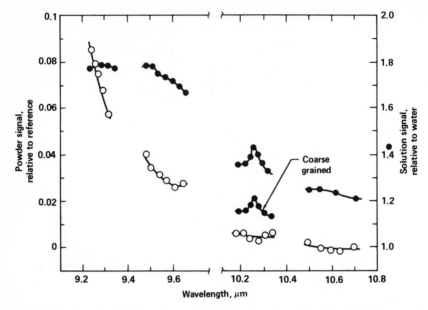

Figure 14.15 Spectra of $(NH_4)_2SO_4$ as a fine-grained powder (●) and in 58.5-g/kg aqueous solution (○), with the feature at 10.26 μm shown for a coarse-grained powder sample as well. (Reproduced by permission from Nordal and Kanstad, 1977.)

surface/volume ratios because of the significant heat flow from the sample to the gas. In the case of liquids and bulk solids that have low surface/volume ratios, piezoelectric detection is often more efficient.

Photoelastic generation of ultrasonic waves in solids predates the recent development of photoacoustic spectroscopy. White (1963) first demonstrated that ultrasonic waves could be produced in solids through transient surface heating of samples by the absorption of pulses of electromagnetic radiation. Since then, there have been many experiments on the generation of high-frequency acoustic signals in both solids and liquids, usually through the use of pulsed lasers.

The first photoacoustic spectroscopic experiments with piezoelectric transducers were perfomed by Hordvik and Schlossberg (1977), who used the technique to study weak absorptions in high-power laser windows. A schematic of their experimental arrangement is shown in Figure 14.16. One of the transducers was attached to the sample with epoxy. In addition to the attached transducer, another transducer was placed symmetrically with respect to the beam and located very close to, but not in acoustical contact with, the sample. The purpose of the latter transducer was to measure the

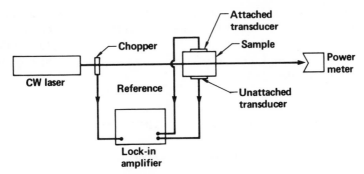

Figure 14.16 Experimental arrangement for a piezoelectric measurement of low optical absorption in laser window materials. (Reproduced by permission from Hordvik and Schlossberg, 1977.)

effect of any radiation scattered onto the detectors, so that, when necessary, the signal measured with the attached transducer could be corrected for the effect of scattered radiation.

Using this arrangement, Hordvik and Schlossberg measured absorption coefficients in the range of 10^{-4} to 10^{-5} cm^{-1} with laser powers of less than 1 W. With samples having lower scatter, they estimate that 10^{-6} cm^{-1} absorption coefficients could be measured with laser powers of 1–10 W. Calibrations were performed by measuring at a wavelength of reasonably strong absorption and comparing with conventional spectrophotometric results.

In another experiment Hordvik and Skolnik (1977) demonstrated that they could distinguish between surface and bulk absorptions in these "transparent" laser windows. This capability results from the fact that an acoustic wave generated at the surface of the sample acts approximately as a spherical wave having components that interact differently with the transducer located on a side than does a bulk acoustic wave, which propagates with nearly radial symmetry. The piezoelectric signals from surface and bulk absorption thus have different phases and can therefore be distinguished. Hordvik and Skolnik found that all the laser window materials they studied had surface absorption losses comparable to or greater than bulk losses. In addition, they found that while some of the samples displayed intrinsic multiphonon bulk absorption, others did not, indicating that their bulk absorptions were impurity dominated.

Farrow et al. (1978) demonstrated that photoacoustic spectra of both bulk and powdered samples could also be obtained with piezoelectric detection methods.

Several experiments have been performed with liquid samples contained in piezoelectric cells. Some of these experiments are described in Chapter 13. Other experiments deal with fluorescence and quantum efficiency studies in liquids, and these are discussed in Chapter 19, which deals with deexcitation studies.

14.9   NOVEL SPECTROSCOPIES

14.9.1   Dichroism Spectroscopy

In conventional photoacoustic spectroscopy, the periodic heating of the sample results from modulation of the intensity of the incident radiation. One can obtain an equivalent effect by keeping the intensity constant, but modulating the wavelength of the incident radiation. Another modulation technique has been employed by Fournier et al. (1978), who kept the light intensity constant but modulated its polarization and thus were able to investigate materials that are optically active. A block diagram of their apparatus is shown in Figure 14.17. The linearly polarized light of the CW dye laser is sequentially set right- or left- circular by a Pockels modulator

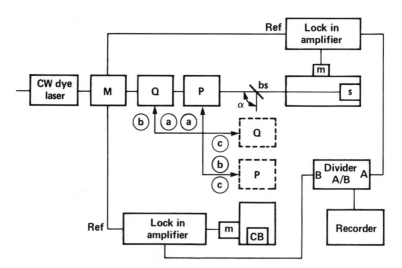

Figure 14.17 Experimental arrangement for a photoacoustic dichroism experiments. Absorption measurements with P and Q in position a. Linear dichroism measurements with P and Q in position b. Circular dichroism measurements with P and Q in position c. (M) microphone, S sample, CB carbon black, (bs)-beam splitter. (Reproduced by permission from Fournier et al,. 1978.)

M. These circular polarizations may be transformed into a set a linearly orthogonal polarizations by an achromatic quarter-wave plate Q. Using a polarizer P one may obtain an intensity-modulated beam. By switching P and (or) Q from the sample beam to the reference beam, it is possible to measure a normalized absorption (PAS) spectrum, a linear dichroism signal (b position), or a circular dichroism signal (c position).

Figure 14.18 shows the wavelength dependence for a 2-mm optically active crystal of $Nd_2(MoO_4)_3$. The bottom curves represent the normal PAS spectra for π and σ linearly polarized light. The upper curve represents the linear dichroic PAS spectrum obtained by modulating between π and σ polarizations. The upper curve is essentially identical to what one would expect to obtain by taking the difference $\beta_\pi - \beta_\sigma$ from the bottom curves.

The axial absorption and magnetic circular dichroism ($MCD = \beta_{\pi+} - \beta_{\sigma-}$) photoacoustic spectrum exhibits, as expected, mainly S-shaped terms associated with the shift of the σ^+ and σ^- transitions due to the ground- and excited-state Zeeman splittings. For a field of 7 kOe, Fournier et al. found that they could measure a dichroic $\nabla\beta/\beta$ of about 10^{-3}.

Dichroic PAS thus appears feasible and provides data comparable to conventional optical dichroic techniques. However, dichroic PAS can be used to measure natural or induced optical anisotropy of a large number of materials (crystals, polymers, or biological samples) that, because of opacity or light scattering, cannot be studied by conventional spectroscopies.

14.9.2   Fourier-Transform PAS

At present photoacoustic spectrometers using wide-band tunable sources operate only into the near infrared. However, most of the important

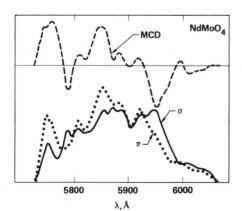

Figure 14.18 (Lower curve) π and σ photoacoustic linear dichroism spectra of $Nd_2(MoO_4)_3$ crystal. (Upper curve) photoacoustic magnetic circular dichroism spectrum of the $Nd_2(MoO_4)_3$ crystal. (Reproduced by permission from Fournier et al., 1978.)

characteristic spectral details of compounds occur in the midinfrared (5–15 μm). Since conventional broad-band infrared sources are too weak for photoacoustic experiments using conventional monochromaters, PAS studies in the mid-IR have so far been performed only with narrow-band laser sources.

Another alternative is to use transform spectroscopy, and a Fourier transform PAS spectrometer has been designed by Farrow et al. (1978b). A Michelson interferometer has two advantages over a conventional monochromator

1. Data in an interferometer are collected from all spectral frequencies simultaneously throughout the duration of the measurement. This allows a much higher signal/noise ratio (Fellgett's advantage) than a conventional monochromator (Griffiths et al., 1977).
2. An interferometer has a much higher optical throughput than a monochrometer (Jacquinot's advantage) (Griffiths et al., 1977).

Figure 14.19 is a block diagram of the Fourier-transform PAS spectrometer designed by Farrow et al. Modulation of the output light from the interferometer was achieved by driving one mirror sinusoidally. In addition, this mirror was also translated, as in a conventional interferometer,

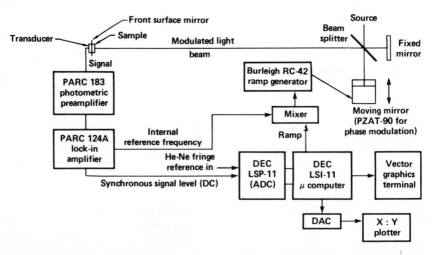

Figure 14.19 Schematic of Fourier-transform visible photoacoustic spectrometer. The Michelson interferometer is in the top right-hand corner. (Reproduced by permission from Farrow et al., 1978b.)

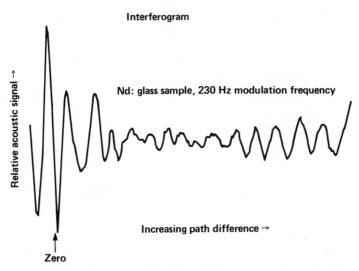

Interferogram

Relative acoustic signal ↑

Nd: glass sample, 230 Hz modulation frequency

Increasing path difference →

Zero

Figure 14.20 Photoacoustic interferogram of a Nd/glass sample. (Reproduced by permission from Farrow et al., 1978b.)

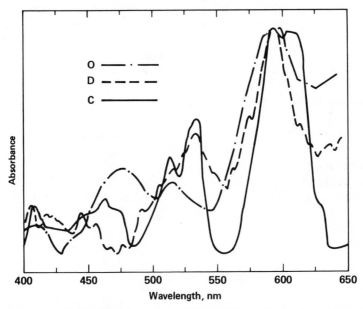

O ——— · ———
D ——— ———
C ———————

Absorbance

400   450   500   550   600   650

Wavelength, nm

Figure 14.21 Absorption spectra of a Nd–glass sample. Curve O is calculated from the interferogram of Figure 14.20. curve D is a conventional photoacoustic absorption spectrum. Curve C is a typical absorption spectrum taken from the literature. (Reproduced by permission from Farrow et al., 1978b.)

189

by mixing a ramp voltage with the sinusoidal voltage to achieve a phase-modulated interferogram. To minimize phase complications, a piezoelectric detection method for the photoacoustic signal was used.

The experiment was performed in the visisble 400–650 nm region and a Nd–glass sample was used. The light source was a 100-W tungsten–iodide lamp. The interferogram is shown in Figure 14.20, and the resulting spectrum is shown as curve a in Figure 14.21. Curve b is a normal PAS spectrum obtained with a conventional dispersive monochromater, and curve c is an absorption spectrum taken from the literature. Spectrum a was obtained in 4 min, while spectrum b took 90 min with a 450-W lamp.

14.9.3 Nanosecond Spectroscopy

Photoacoustic spectroscopy is generally not considered suitable for fast measurements. The gas-microphone system is not only limited by the speed of sound in the sample and gas, but even more so by the slow response of the microphone ($< 20 \mu$sec). Even a piezoelectric system generally responds no faster than 0.1 μsec.

Rockley and Devlin (1977) have shown, however, that nanosecond resolution is achievable with the slow gas-microphone system under certain conditions. Using two pulsed dye lasers of only slightly different wavelengths, with 6 nsec pulse width, Rockley and Devlin found that a marked enhancement of the PAS signal was obtained from a Rose Bengal dye when the two laser beams were directed at the same spot in the sample at the same time. They explained this result by absorption not only from the ground state to the first excited state, but by simultaneous absorption from the first excited state into higher levels. The lifetime of the first excited state is about 0.9 nsec. Thus, provided the delay between the two lasers pulses is less than 0.9 nsec, an enhanced PAS signal results from excited state absorption.

This technique can therefore be used to measure the absorption spectra and time-resolved absorption spectra of transient excited states by varying the wavelength of one of the lasers and by incorporating time delays between the two pulses.

14.10 CONCLUSIONS

Spectroscopy, the study of the interaction of energy with mattter, is the prime occupation of most physicists and chemists. It is therefore not surprising that the photoacoustic effect has already been used in a wide variety of physical and chemical studies where it has proved its worth. It will undoubtedly become a familiar tool in physical and chemical laboratories in the near future, not only to extend the power of spectroscopic analysis to hitherto unmanageable materials, but also to be used with any and all materials because of some of its truly unique capabilities.

REFERENCES

Adams, M. J., Beadle, B. C., King, A. A., and Kirkbright, G. F. (1976). *Analyst* **101**, 553.

DeGennes, P. G. (1974). *The Physics of Liquid Crystals* Oxford University Press, London and New York.

Eaton, H. E., Anton, D. R., and Stuart, J. D. (1977). *Spectrosc. Lett.* **10**, 847.

Farrow, M. M., Burnham, R. K., Auzanneau, M., Olsen, S. L., Purdie, N., and Eyring, E. M. (1978a). *Appl. Opt.* **17**, 1093.

Farrow, M. M., Brunham, R. K., and Eyring, E. M. (1978b). *Appl. Phys. Lett.* **33**, 735.

Fernelius, N. C. (1978). *Appl. Spectrosc.* **32**, 554.

Fournier, D., Boccara, A. C., and Badoz, J. (1978). *Appl. Phys. Lett.* **32**, 640.

Griffiths, P. R., Sloane, H. J., and Hannah, R. W. (1977). *Appl. Spectrosc.* **31**, 485.

Hordvik, A., and Schlossberg, H. (1977). *Appl. Opt.* **16**, 101.

Hordvik, A., and Skolnik, L. (1977). *Appl. Opt.* **16**, 2919.

Lewis, N. S., Mann, K. R., Gordon, J. G., and Gray, H. B. (1976). *J. Am. Chem. Soc.* **98**, 7461.

Lin, J. W., and Dudek, L. P. (1979). *Anal. Chem.* **51**, 1627.

McClure, D. S. (1963). *J. Chem. Phys.* **38**, 2289.

Nelson, D. F., Johnson, L. F., and Gershenzon M. (1964). *Phys. Rev.* **135**, A1399.

Nordal, P.-E., and Kanstad, S. O. (1977). *Opt. Commun.* **22**, 185.

Pankove, J. I. (1971). *Optical Processes in Semiconductors*, Prentice-Hall, New York.

Rockley, M. G., and Devlin, J. P. (1977). *Appl. Phys. Lett.* **31**, 24.

Rosencwaig, A. (1973). *Opt. Commun.* **7**, 305.

Rosencwaig, A. (1975a). *Anal. Chem.* **47**, 592A.

Rosencwaig, A. (1975b). *Phys. Today* **28**, (9), 23.

Rosencwaig, A., Ginsberg, A. P., and Koepke, J. W. (1976). *Inorg. Chem.* **15**, 2540.

Somoano, R. B. (1978). *Angew. Chem. Int. Ed.* (*Engl.*) **17**, 234.

Tandon, S. P., and Gupta, J. P. (1970). *Phys. Status Solidi* **3**, 329.

White, R. M. (1963). *J. Appl. Phys.* **34**, 3559.

Williams, R., Hsu, C. H., Cuellar, E., Gordon, J., Samson, S., Hadek V., and Somoano, R. (1977). *Am. Chem. Soc. Div. Org. Coat. Plast. Chem.*, Paper **37**, 316.

CHEMICAL STUDIES

15.1 INTRODUCTION

Many chemical studies performed with photoacoustics either are analytical in nature, wherein the aim is in detecting and identifying trace elements as described in Chapter 13, or they fall into the broad category of spectroscopic studies, such as those discussed in Chapter 14. In this chapter, we consider the application of photoacoustics to studies in the important fields of catalysis and chemical reactions.

15.2 CATALYSIS AND CHEMICAL REACTIONS

Photoacoustics is ideally suited to catalytic studies since catalytic substances are by their very nature difficult to investigate by conventional spectroscopic means. These difficulties arise from the fact that in heterogeneous catalysis, the catalyst is often in the form of a fine powder.

In Figure 15.1a we show data obtained from an experiment on the inorganic insulator system $CoMoO_4$ (Rosencwaig, 1975). This experiment was performed in the hope of acquiring a better understanding of the hydrodesulfurization catalytic action of $CoMoO_4$ supported on alumina (Lipsch and Schmit, 1969). Both the high-temperature, β-$CoMoO_4$, and low-temperature, α-$CoMoO_4$, phases are available only as fine precipitates and thus their optical spectra are not readily obtainable by conventional techniques. The photoacoustic spectra shown in Figure 15.1a indicate that the β-$CoMoO_4$ has a charge-transfer band similar to that seen in the parent MoO_3, while the charge-transfer band in the α-$CoMoO_4$ has shifted noticeably to lower energy. Figure 15.1b is a more detailed study of the crystal field bands of the Co^{2+} ions in $CoSO_4 \cdot H_2O$ (octahedral coordination) in α- and β-$CoMoO_4$, and in $CoAl_2O_4$ (tetrahedral coordination). These PAS spectra indicate that the Co^{2+} ions in both the α- and β-$CoMoO_4$ are in a distorted octahedral coordination, and that there is no significant difference in the Co^{2+} d-electron configuration between the two $CoMoO_4$ phases. This information then implies that the observed difference in catalytic activity between the α- and β-phases (the α-phase is

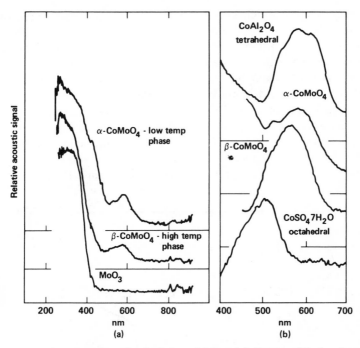

Figure 15.1 A photoacoustic study of cobalt molybdate. (*a*) Spectra of the low-temperature alpha and the high-temperature beta forms of $CoMoO_4$. (*b*) Spectra of the Co^{2+} crystal-field bands on hydrated cobalt sulfate (octahedral cubic), the two forms of cobalt molybdate, and cobalt aluminate (tetrahedral). (Reproduced by permission from Rosencwaig, 1975.)

catalytically more active than the β-phase) cannot be attributed solely to differences in the localized 3-d electron configuration.

However, Figure 15.1*a* does provide another possible explanation for this difference. We note that the charge-transfer band edge of the α-$CoMoO_4$ is at a considerably lower energy than that of the β-$CoMoO_4$. If this were the only significant difference between the two phases, then, based on these data alone, we would predict that the α-$CoMoO_4$ would be catalytically more active than the β-$CoMoO_4$, since it requires less energy to excite electrons into the "mobile" charge-transfer state in α-$CoMoO_4$ than in β-$CoMoO_4$. Since this prediction is in agreement with experimental data, the photoacoustic study indicates that the possibility of differences in catalytic activity arising from differences in ligand-electron configuration should be investigated further.

Another example in the field of catalysis involves the investigation of the reaction of transition metal complexes with polymeric ligands to form anchored catalysts, which have been used to catalyze hydrogenation and hydroformulation of olefins. PAS has been used in the investigation of the electronic structure of these metal–polymer complexes to elucidate chemical processes and structure–reactivity relationships. An example of a PAS study of a model catalytic system involving a well-characterized compound, tungsten hexacarbonyl $W(CO)_6$, is shown in Figure 15.2.

The PAS spectrum of $W(CO)_6$ is shown both before and after photochemical reaction with polyvinylpyridine (Somoano, 1978). Figure 15.2*a* shows the PAS spectrum of $W(CO)_6$ and reveals the singlet–triplet ligand field transition at 350 nm, as well as the corresponding singlet–singlet transition at 310 nm (Wrighton et al., 1976). Figure 15.2*b* shows PAS

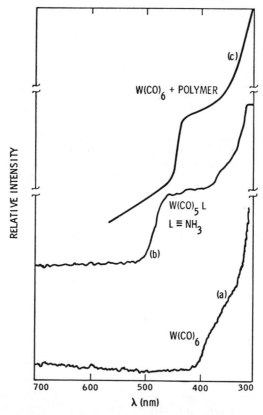

Figure 15.2 Photoacoustic spectra of tungsten hexacarbonyl, $W(CO)_6$ complexes. (*a*) $W(CO)_6$, (*b*) $W(CO)_5NH_3$, (*c*) $W(CO)_6$+polyvinylpyridine. The complexes in *b* and *c* are photoproducts. (Reproduced by permission from Somoano, 1978.)

spectrum of $W(CO)_5L$, where the ligand L is NH_3. This compound is formed upon irradiation of $W(CO)_6$ in the presence of the ligand and is shown to reveal the effect of ligand substitution in $W(CO)_6$. The characteristic singlet–triplet and singlet–singlet transitions at 457 and 416 nm, respectively, of $W(CO)_5NH_3$ are observed in the form of a broad absorption band from 375 to 475 nm.

Figure 15.2c shows the PAS spectrum of the photoproduct of $W(CO)_6$ and polyvinylpyridine and clearly reveals the $W(CO)_5$–pyridine absorption band similar to that seen in Figure 15.2b. The polymer-anchored $W(CO)_5$ species has been observed to catalyze olefin isomerization and hydrogenation. Thus PAS may be used to study and characterize solid-state photoproducts where conventional optical and structural (i.e., x-ray) techniques would be totally inadequate.

An example of a less-well-understood, but potentially useful, polymer-anchored catalyst involves the transition metal cluster complex tetra-rhodium dodecacarbonyl, $Rh_4(CO)_{12}$. Figure 15.3 shows the photoacoustic spectra: (a) $Rh_4(CO)_{12}$, (b) the photoproduct of the reaction of $Rh_4(CO)_{12}$ solution with polyvinylpyridine, and (c) the thermal product (i.e., the same as b but not exposed to UV radiation) (Somoano, 1978). Both of the polymer–metal complexes, b and c, are found to be hydroformulation catalysts, but the phototriggered catalyst, b, is considerably more active. This extra activity is associated with the absorption band (shoulder) at 600 nm in Figure15.3b since the less active thermal catalyst lacks this absorption. On prolonged exposure to air, the phototriggered catalyst decomposes, as indicated by the loss of the 600-nm absorption. Its PAS spectrum then resembles that of the thermal catalyst and exhibits correspondingly less catalytic activity (Somoano et al, 1978).

Another use of photacoustic spectroscopy in this field has been to detect the degree of reduction of metals reacted on polymeric substrates. By using PAS to monitor the reaction of, say, chloroplatinate, K_2PtCl_4, with polyvinylpyridine, one may easily tell when the chloroplatinate has been fully reduced to platinum metal. The use of PAS in catalysis studies is just starting, but there is every indication that it will prove to be a valuable tool.

PAS has also been used for the study of chemically treated controlled pore glasses. These materials are growing in importance in the fields of chemical synthesis and analysis. Functional groups such as metal chelation reagents, acid–base indicators, and enzymes can be attached to the surface of the glass and used in a variety of chemical reactions. However, techniques for definitive characterization of the attached functional groups are rather primitive and ineffective. The photoacoustic technique has been

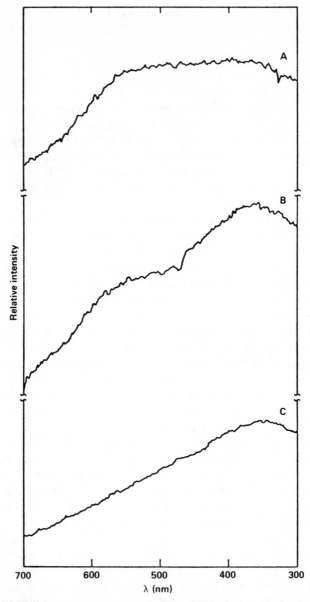

Figure 15.3 Photoacoustic spectra of tetrarhodium dodecacarbonyl, $Rh_4(CO)_{12}$, polymer complexes. (a) $Rh_4(CO)_{12}$, (b) the photoproduct of the reaction of $Rh_4(CO)_{12}$ solution with polyvinylpyridine, and (c) same as b, but not irradiated with UV light (i.e., a thermal product). (Reproduced by permission Somoano, 1978.)

used to determine the presence of these groups and to monitor subsequent chemical reactions and identify photochemical degradation processes.

It is clear that the applications of photoacoustic spectroscopy in the fields of catalysis and chemical reactions have significant potential. For example, it is possible to construct a reaction chamber that is also a photoacoustic cell. Even in the presence of an ongoing exothermic reaction, PAS spectra can be obtained at a solid–gas interface, and thus dynamic studies of catalysis and other chemical reactions can be performed. Similar studies could also be conducted on solid–liquid interfaces using a liquid photoacoustic cell.

REFERENCES

Lipsch, J. M. J. G., and Schmit, G. C. A. (1969). *J. Catal.* **15**, 163.

Rosencwaig, A. (1975). *Anal. Chem.* **47**, 592A.

Somoano, R. B. (1978). *Angew. Chem. Int. Ed.* (*Engl.*) **17**, 234.

Somoano, R. B., Gupta, A., Volksen, W., Rembaum, A., and Williams, R. (1978). In *Organometallic Polymers* (C. E. Carraher, Jr., J. E. Sheats, and C. U. Pittman, Jr., Eds.), Academic Press, New York.

Wrighton, M. W., Morse, D. L., Gray, H. B., and Ottesen, K. K. (1976). *J. Am. Chem. Soc.* **98**, 1111.

CHAPTER

16

SURFACE STUDIES

16.1 INTRODUCTION

Photoacoustic spectroscopy can be used to great advantage in the study of adsorbed and chemisorbed molecular species and compounds on the surface of metals, semiconductors, and even insulators. Such studies can be performed at any wavelength, providing that the substrate one is dealing with is either nonabsorbing or highly reflecting at this wavelength. Under either of these conditions, the PAS experiments give the optical absorption spectra of the adsorbed or chemisorbed compounds on the surface of the substrate.

16.2 SURFACE STUDIES IN THE UV–VISIBLE

The first indication of the feasibility for using photoacoustic spectroscopy for surface studies came from an experiment performed with thin layer chromatography (Rosencwaig and Hall, 1975). Thin layer chromatography, or TLC, is a widely used and highly effective technique for the separation of mixtures into their constituents (Randerath, 1966). This technique is of considerable importance in the chemical, biological, and medical fields. The identification of the TLC-separated compounds directly on the TLC plates can, however, be a fairly difficult procedure, particularly if reagent chemistry is inappropriate. Conventional spectroscopic techniques are unsuitable because of the opacity and light-scattering properties of the silica gel adsorbent on the TLC plates. Photoacoustic spectroscopy offers a simple and highly sensitive means for performing nondestructive compound identification directly on the TLC plates.

In Figure 16.1 we show, on the left, the PAS spectra taken on five different compounds that were separated and developed on TLC plates. These PAS spectra were taken directly on the plates themselves and were run in the ultraviolet region of 200–400 nm. The five compounds starting from the top of Figure 16.1 are (*I*) p-nitroaniline, (*II*) benzylidene acetone, (*III*) salicylaldehyde, (*IV*) 1-tetralone and (*V*) fluorenone. On

198

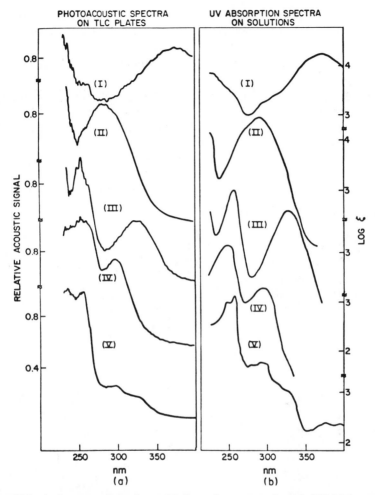

Figure 16.1 A photoacoustic study on thin-layer chromatography. (*a*) UV PAS spectra of five compounds: (*I*) *p*-nitroaniline, (*II*) benzylidene-acetone, (*III*) salicylaldehyde, (*IV*) 1-tetralone, and (*V*) fluorenone. These spectra were taken directly on the thin-layer chromatography plates. (*b*) UV absorption spectra of the same five compounds in solution. (Reproduced by permission from Rosencwaig and Hall, 1975.)

the right are shown, for the sake of comparison, the published UV absorption spectra of these compounds in solution (Perkampus et al., 1966–1971). The strong similarity between the photoacoustic spectra and the optical absorption spectra permits a rapid and unambiguous identification of the compounds.

To test the sensitivity of the PAS technique in this application the investigators performed the experiment shown in Figure 16.2. Here are the PAS spectra taken directly on a TLC plate on which benzylidene acetone spots of different concentrations were developed. The spectra were taken

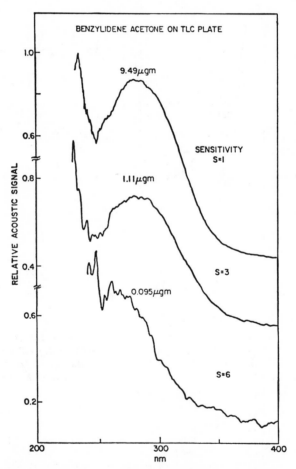

Figure 16.2 Photoacoustic spectra on spots of benzylidene acetone on a TLC plate. The spots were developed from starting drops containing 9.49, 1.11, and 0.095 µg of benzylidene acetone. (Reproduced by permission from Rosencwaig and Hall, 1975.)

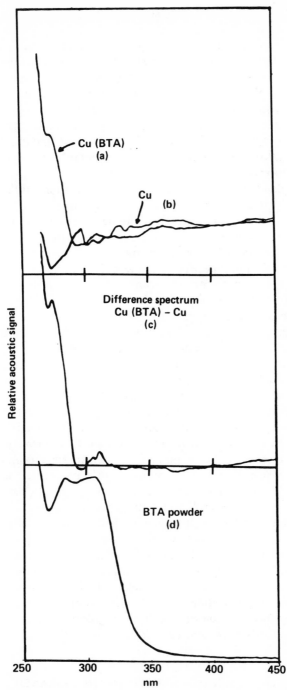

Figure 16.3 A PAS surface study. Spectrum *a* is a photoacoustic spectrum of a highly polished copper surface treated with benzotriazole (BTA). Spectrum *b* is of an untreated surface. Spectrum *c* is the difference spectrum, showing the absorption bands of the monolayer of BTA. Spectrum *d* is of bulk benzotriazole powder. (Reproduced by permission from Rosencwaig, 1975.)

on spots developed from starting drops containing $\sim$ 10, 1, and 0.1μg of benzylidene acetone. The spectra were taken on the developed spots, and these can be expected to have an even smaller amount of material than the starting drops. Nevertheless, we note that even for the case of 0.1 μg, the main absorption band of the benzylidene acetone is visible. Knowing the amount of material present in this spot ($<$ 1μg), the size of the spot ($\sim$ 0.3 cm^2 in area), and the molecular weight of the compound (146), it was estimated that roughly no more than one monolayer of the compound was present in this spot.

An example of the use of PAS in surface passivation studies of metals is shown in Figure 16.3 (Rosencwaig, 1975). Spectrum a at the top is of a copper surface that has been treated with benzotriazole, a known passivating agent. Spectrum b is of an identical but untreated copper surface. Spectrum a differs markedly from spectrum b only in the wavelength region below 300 nm. Spectrum c is the difference spectrum produced by subtracting spectrum b from spectrum a. Spectrum c thus represents the spectrum of the chemisorbed monolayer of benzotriazole. Spectrum d at the bottom is a PAS spectrum of some benzotriazole powder. We note that spectrum c is quite different than spectrum d, indicating that the chemisorbed benzotriazole has undergone significant structural change. It may be possible to establish, from such spectra, what changes have occurred to the benzotriazole upon chemisorption onto the copper surface, and thus to understand how this compound passivates the surface of copper metal.

These experiments indicate the possibility that under certain conditions of the substrate (low optical absorption or high reflectivity), the PAS technique may well be sensitive enough to detect and identify a monolayer of adsorbed or chemisorbed compound. Other experiments on both metallic and nonmetallic surfaces have confirmed that monolayer detectability is achievable. The use of photoacoustic spectroscopy for such surface studies, particularly with high-resolution light sources such as tunable dye or infrared lasers, can lead to fundamental understandings of surface oxidation and reduction processes under a variety of conditions, and also to further knowledge about catalytic activity on solid surfaces. Other surface studies would include PAS studies of organic compounds and inorganic oxides deposited on the surface of metals, semiconductors, and polymers for purposes of passivation. Such studies would yield data about the structure, valence, complexing, and so on of the deposited compound, information that is at present very difficult to obtain nondestructively.

16.3   SURFACE STUDIES IN THE INFRARED

The PAS experiments performed in the visible region of the electromagnetic spectrum demonstrate the feasibility of studying surfaces and adsorbed species by this technique. It would, however, be more useful to be able to obtain spectra of adsorbed species in the mid-infrared region, since the information there is much more detailed and specific. Most photo-acoustic spectrometers, including the commercial systems, can be operated into the near-IR up to 2.5–3.0 μm. Low and Parodi (1978) have shown that it is possible to perform PAS surface studies with a conventional 2000° K IR source to 4.5 μm.

These researchers investigated solid chromia alumina catalyst pellets before and after degassing and silica powder under various chemical conditions, as shown in Figure 16.4. Trace A of the untreated silica powder shows merely the broad and asymmetric absorption characteristics of surface silanols and sorbed water. After the sample has been mildly degassed and treated with $(CH_3)_3CH$, the O-H absorption decreases in intensity and a pair of bands near 2960 and 2860 cm^{-1}, characteristic of surface Si-OCH$_3$ becomes evident, as seen in curve B. When the sample is exposed to HSiCl$_3$ vapor at 300–400° C, further changes occur, the Si-OH band shows a further decline and splittings, the 3740 cm^{-1} band of "isolated" or "free" silanols becomes more discernible, and the 2270 cm^{-1}

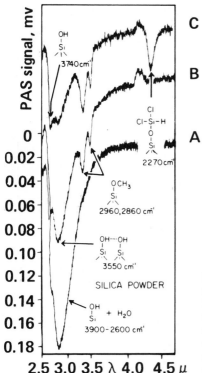

Figure 16.4 Photoacoustic spectra of surface species on silica. (A) Fresh silica powder after a 100°C heating in air to remove water. (B) After treatment with $(CH_3O)CH$, causing a partial esterification of the surface. (C) After treatment with HSiCl$_3$, causing a partial silanization of the surface. (Reproduced by permission from Low and Parodi, 1978.)

Si-H stretching band characteristic of the chemisorbed silane appears
(Little, 1966; Kiselef and Lygin, 1975).

Although this is only a preliminary experiment and the spectral resolution obtained is rather poor in comparison to absorption spectra recorded
with high-quality dispersion instruments, it does illustrate the potential of
this technique, and it must be kept in mind that such experiments can be
performed even in the presence of intense light scattering from the substrate material. In addition, if one can use a Fourier-transform PAS
spectrometer, similar to the one described in Chapter 14, then surface
studies beyond 4.5 μm become practical as well.

In the meantime, it is possible to perform some PAS surface studies in
the mid-IR by means of lasers such as the CO_2 laser. Nordal and Kanstad

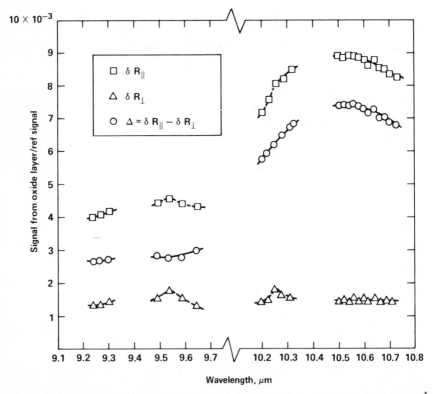

Figure 16.5 Photoacoustic spectra of an aluminum oxide film anodized to 10 V ($\sim$160 Å
thick), showing δR_{11}, $\delta R_{\perp}$ and $\Delta = \delta R_{11} - \delta R_{\perp}$. (Reproduced by permission from Nordal
and Kanstad, 1978.)

(1978) studied thin oxide films on aluminum by this method. Such studies are conventionally performed by reflection–absorption spectroscopy using light polarized either parallel to the plane of incidence ($\delta R_{\parallel}$) or perpendicular to the plane of incidence ($\delta R_{\perp}$). Because the change in refelction due to the small absorption that occurs in the film is exceedingly small, these experiments are quite difficult. On the other hand, since only the absorbed light produces a photoacoustic signal, PAS can improve and simplify such measurements.

Samples of oxide on aluminum were prepared by anodization, which produced a well-defined (nonporous) oxide film of thickness ~ 14 Å/V (Plumb, 1958).

In Figure 16.5 are shown the PAS spectra of $\delta R_{\parallel}$ (upper trace) and $\delta R_{\perp}$ (lower trace) for a sample anodized to 10 V (~ 160 Å thick). The middle trace gives the difference curve $\Delta = \delta R_{\parallel} - \delta R_{\perp}$, where δR represents reflection loss (or absorption due to the film). A strong absorption maximum is

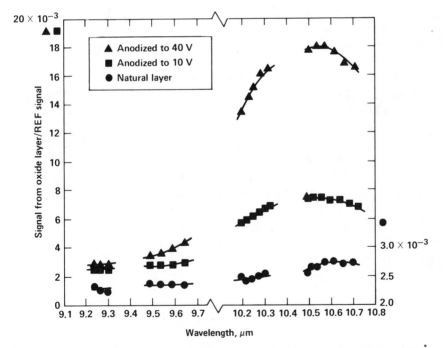

Figure 16.6 Photoacoustic Δ spectra of aluminum oxide films anodized to 40 V (~ 580 Å) and 10 V (~ 160 Å), respectively, and the spectrum of a natural (air) oxide layer (~ 15–30 Å). Note the change of ordinate scale for the last spectrum. (Reproduced by permission from Nordal and Kanstad, 1978.)

clearly evident at 10.55 μm. Its origin is generally identified as the longitudinal mode of the Al-O stretching vibration (Maeland et al., 1974). Only parallel polarized radiation couples to these vibrations.

Figure 16.6 shows the spectra of three samples: one sample with a natural (air) oxide layer ($\sim$15–30 Å thick), the preceding sample anodized to 10 V, and one sample anodized to 40 V ($\sim$580 Å thick). As we see, the absorption peak due to the oxide layer increases linearly with layer thickness. In fact, Nordal and Kanstad estimate that with laser intensities of a few watts, it would be possible to measure absorptions of layers, with characteristics similar to those of aluminum oxide, down to an average thickness of 10^{-3} monolayers. This is comparable to what is attainable with electron energy loss spectroscopy, but with the superior resolution of optical spectroscopy.

REFERENCES

Kiselef, A. V., and Lygin, V. I. (1975). *Infrared Spectra of Surface Compounds*, Wiley, New York.

Little, L. H. (1966). *Infrared Spectra of Adsorbed Species*, Academic Press, New York.

Low, M. J. D., and Parodi, G. A. (1978). *Spectrosc. Lett.* **11**, 581.

Maeland, A. J., Rittenhaus, R., Lahar, W., and Romano, P. V. (1974). *Thin Solid Films* **21**, 67.

Nordal, P. -E., and Kanstad, S. O. (1978). *Opt. Commun.* **24**, 95.

Perkampus, H. H., Sandeman, I., and Timons, C. J. (Eds.) (1966–1971). *DMS Atlas of Organic Compounds*, Vols. I–V, Plenium Press, New York.

Plumb, R. C. (1958). *J. Electrochem. Soc.* **105**, 498.

Randerath, K. (1966). *Thin-Layer Chromatography*, 2nd rev. ed., Academic Press, New York.

Rosencwaig, A. (1975). *Anal. Chem.* **47**, 592A.

Rosencwaig, A., and Hall, S. S. (1975). *Anal. Chem.* **47**, 548.

STUDIES IN BIOLOGY

17.1 INTRODUCTION

One of the most promising areas for the use of the photoacoustic method is in the study of biological systems, for here most of the materials to be studied are often in a form that makes them difficult, if not impossible, to investigate by any other optical technique. Although many biological materials occur naturally in a soluble state, many others are membrane bound or are part of bone or tissue structure. These materials are insoluble and function biologically within a more or less solid matrix. Optical data on these materials are usually difficult to obtain by conventional techniques, since these materials are generally not in a suitable state for conventional transmission spectroscopy, and if solubilized, are often significantly altered. Photoacoustic spectroscopy, through its capability of providing optical data on intact biological matter, even on matter that is optically opaque, holds great promise as both a research and diagnostic tool in biology and medicine.

17.2 Hemoproteins

To illustrate the capabilities of PAS in biology we consider an experiment on cytochrome C (Rosencwaig, 1973), with the results shown in Figure 17.1. The hemoprotein cytochrome C plays an essential role in cellular respiration and has been extensively studied (Margoliash and Schijter, 1966). Since it is readily soluble in water, its optical absorption spectrum is well known. Figure 17.1a shows the optical absorption spectrum of the oxidized and reduced forms of cytochrome C in aqueous solution obtained with a conventional spectrophotometer. Figure 17.1b shows the PAS spectra of oxidized and reduced cytochrome C in lyophilized or powder form.

We note that the photoacoustic spectra are qualitatively very similar to the optical absorption spectra. In particular, all the differences between the oxidized and reduced forms visible in the absorption spectra are also visible in the photoacoustic spectra. This experiment indicates that it is possible with photoacoustic spectroscopy to obtain spectra on biological material in solid form that are comparable to those obtained in solution.

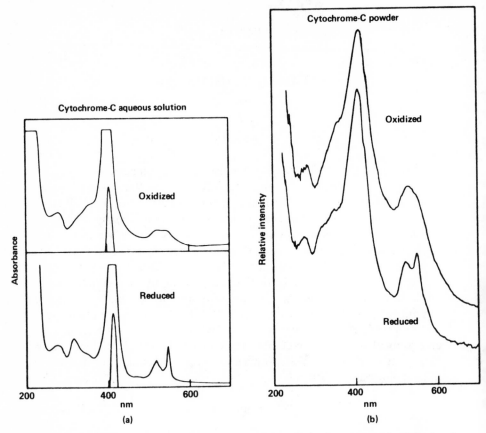

Figure 17.1 Study of cytochrome C. (*a*) The optical absorption spectra of oxidized and reduced cytochrome C in aqueous solution. (*b*) The photoacoustic spectra of oxidized and reduced cytochrome C in solid (powder) form. (Reproduced by permission from Rosencwaig, 1973. Copyright 1973 by the American Association for the Advancement of Science.)

This capability becomes extremely important when the compound to be studied is insoluble.

The oxygen-carrying protein of red blood cells, hemoglobin (Antonini and Brunori, 1971), also displays a strong and characteristic heme absorption spectrum. Conventional absorption spectroscopy on whole blood, even when diluted in a suitable buffer, does not produce satisfactory results. The inadequacy of the conventional technique arises from the strong light-scattering properties of whole blood, due primarily to the presence of the other protein and lipid material in the plasma and to the

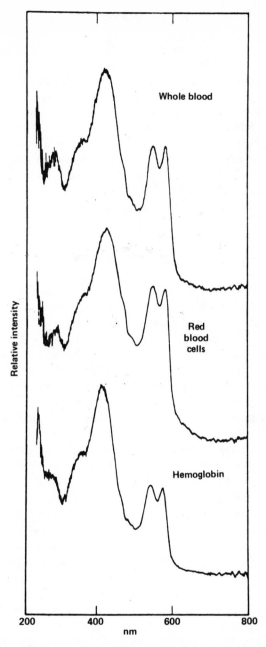

Figure 17.2 Photoacoustic spectra of smears of whole blood, red blood cells, and hemoglobin. All three spectra clearly show the band structure of oxyhemoglobin. (Reproduced by permission from Rosencwaig, 1973. Copyright 1973 by the American Association for the Advancement of Science.)

red blood cells. In the conventional process, one first extracts the hemo-globin from the whole blood by centrifuge techniques and then obtains an absorption spectrum of an aqueous solution of the extracted hemoglobin. The PAS spectrum of a smear of whole blood in Figure 17.2 exhibits the characteristic spectrum of oxyhemoglobin as clearly as in the PAS spec-trum of red blood cells and even of the extracted hemoglobin itself (Rosencwaig, 1973). The presence of the other protein and lipid material in whole blood that causes so much difficulty in conventional spectroscopy from light scattering creates no problem in photoacoustic spectroscopy. Thus it is now possible to study hemoglobin directly in whole blood, that is, *in situ*, without resort to extraction procedures.

17.3 PLANT MATTER

The unique capabilities of photoacoustic spectroscopy enable it to be used to obtain optical absorption data on even more complex biological sys-tems, such as green leaf and other plant matter. A PAS spectrum on an intact green leaf in Figure 17.3 (Rosencwaig, 1978) clearly shows all the optical characteristics of the chloroplasts in the leaf matter—the Soret band at 420 nm, the carotenoid band structure between 450 and 550 nm, and the chlorophyll band between 600 and 700 nm. PAS can thus be used to study intact plant matter, even living plant matter, and thus obtain valuable information about normal and abnormal plant processes and pathology.

Since we can study intact plant matter, it is clear that photoacoustic spectroscopy can be used as a quick and efficient screening tool in natural products chemistry (Rosencwaig, 1978). Using the PAS spectrum obtained from only a few milligrams of a natural source, such as a plant, animal, and microorganism, one can readily determine the types and relative concentrations of secondary metabolites present, that is, the chemical by-products of the organism's metabolism, compounds that may prove to have important physiological or biomedicinal value. At present such infor-mation can be arrived at only after laborious extraction and analysis procedures requiring many grams of the natural source.

We illustrate the applications of PAS in natural products chemistry with a study on marine algae, a promising source of new and unusual natural products (Scheuer, 1973). The photoacoustic spectra shown in Figure 17.4 were taken on the algae after they had been air dried, using less than 10 mg of material in each case (Rosencwaig, 1975; 1978). Although there is a considerable amount of spectral detail in all the PAS spectra, we were

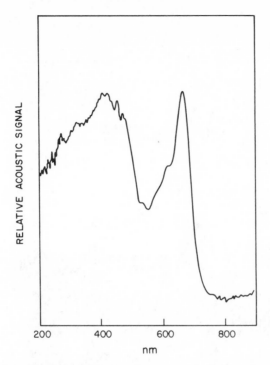

Figure 17.3 A photoacoustic spectrum of an intact green leaf. (Reproduced by permission from Rosencwaig, 1978.)

mainly interested in the region around 320 nm, a region where the sought-after aromatic or highly conjugated secondary metabolites could be expected to have an absorption band. The PAS spectra indicated that only two of the algae would yield any quantity of these aromatic compounds. Conventional extraction and analysis procedures performed by S. S. Hall and his colleagues at Rutgers University have fully confirmed the photoacoustic data. It is clear therefore that photoacoustic spectroscopy can provide, in minutes, information about the compounds present in complex biological systems, and that it is capable of doing this nondestructively, requiring only milligrams of material and no special sample preparation.

Another illustration of the use of photoacoustic spectroscopy in biology is in the study of marine phytoplankton (Ortner and Rosencwaig, 1977), shown in Figures 17.5 and 17.6.

Biochemical characteristics of the algae are being increasingly considered by phytoplankton taxonomists (Norgard et al., 1974). Although

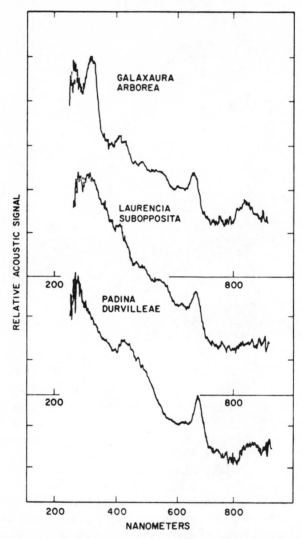

Figure 17.4 Photoacoustic spectra obtained on a few milligrams of dried marine algae. (Reproduced by permission from Rosencwaig, 1978.)

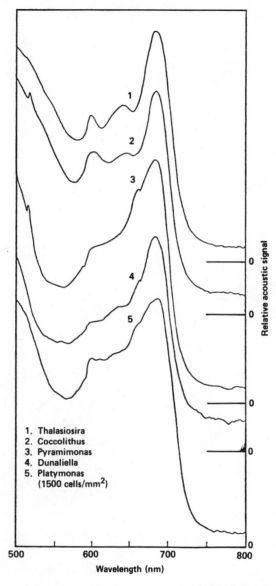

Figure 17.5 Photoacoustic spectra of representative marine phytoplankton. (Reproduced by permission from Ortner and Rosencwaig, 1977.)

213

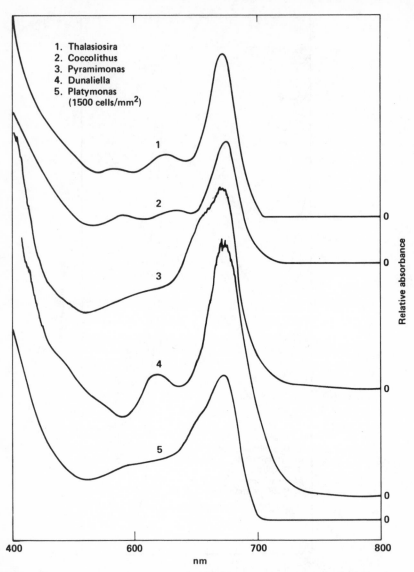

Figure 17.6 Absorption spectra of representative marine phytoplankton. (Reproduced by permission from Ortner and Rosencwaig, 1977.)

complications abound (Jeffrey et al., 1975), photosynthetic pigment composition appears to be a phylogenetically conservative feature (Stewart, 1974). Further, since photosynthetic carbon fixation by microalgae supports the ocean's food chain, the specific photochemical mechanisms by which these cells absorb and transfer light energy are of evident ecological import. Three methods of pigment analysis have been widely employed— chromatography, fluorometry, and spectrophotometry. In general, the resolution and sensitivity of analyses of intact cells has been considerably lower than analysis using solutions of extracted pigment samples (Stewart, 1974). Photoacoustic spectroscopy can, however, be effectively used for analysis of intact cells.

In photoacoustic spectroscopy the signal depends not only on the amount of absorbed optical energy, but also on fluorescent reemission, photochemical transfer, phosphorescence, and thermal deexcitation. Since the PAS signal is a function of thermal deexcitation alone, a PAS spectrum would be closely analogous to an absorption spectrum if fluorescence, photochemical transfer, and phosphorescence were either small, or constant, fractions of the total light absorbed.

In this study the spectra of five phylogenetically disparate algal species were determined in replicate by PAS and by conventional spectophotometric analysis. Monospecific cultures were obtained of *Coccolithus huxleyi*, *Thalassiosira pseudonana*, *Dunaliella tertiolecta*, *Pyramimonas* sp., and *Platymonas* sp. Cells from these cultures were collected on Whatman glass fiber filters and then examined by both PAS and spectrophotometric methods.

In comparing Figures 17.5 and 17.6, it appears that the PAS spectra have more detail than do the absorption spectra. All three chlorophycaen exhibit distinctly separable chlorophyll *b* peaks at 662 nm (Figure 17.5, 3–5). Chlorophyll *c* peaks at 645 nm are clear in both the bacillariophycaen and the haptophycaen (Figure 17.5, 1 and 2). *Coccolithus* and *Thalassiosira* exhibit similar, but unidentified peaks, at 595–600 nm (Figure 17.5, 1 and 2). These features are relatively more distinct in the PAS spectra (compare Figure 17.5 with Figure 17.6). *Platymonas* appears to possess a structured chlorophyll *a* peak at 683–687 nm (Figure 17.5, 5). *Coccolithus* and *Pyramimonas* have sharp unidentified peaks at 518 nm (Figure 17.5, 2 and 3). These features are not observed in the absorption spectra (Figure 17.6). On the other hand, the prominant, but unidentified, 620-nm absorption peak exhibited by *Dunaliella* (Figure 17.6, 4) is not readily discernible in the PAS spectrum (figure 17.5, 4). Overall, PAS spectra roughly parallel absorption spectra. This result is not unexpected, since, according to Sauer and Park (1964), even physically isolated chloroplast fractions, whose photochemical transfer pathways have been chemically blocked, have relatively low fluorescence yields. This result implies that thermal deexcitation rather than fluorescence is preferred when photochemistry is inhibited. Yet, although the absolute energetic shunt represented by fluorescence is rather small, it is suggestive that the fluorescent efficiencies of the different chlorophylls, at least in acetone, are very different (Jeffrey, 1972). If this were true in intracellular pigment systems, it would account for some of the observed differences in the PAS and the absorption spectra. For example, if it were true in whole cells, as it is in acetone, that the fluorescent quantum efficiency of chlorophyll *a* were much greater than the fluorescent quantum efficiency of chlorophyll *b*—0.24 versus 0.09 (Jeffrey, 1972), the PAS signal of the larger chlorophyll *a* peak would be relatively smaller, resulting in greater separation from the overlapping chlorophyll *b* peak. Accurate knowledge of the fluorescent activation spectra of intracellular pigments might also explain the wavelength shifts, relative to spectrophotometry, of the principle chlorophyll peaks identified by PAS. It is equally plausible that the observed spectral differences result from differential heat transfer by the different pigment molecules since they have very different intracellular microenvironments.

These experiments illustrate the usefulness of the photoacoustic technique in the study of intact biological matter. The study on marine phytoplankton also demonstrates that the PAS method may be a more sensitive tool for discriminating between quite similar specimens than conventional absorption spectra. This added sensitivity is due to the dependence of the photoacoustic signal not only on the absorption characteristics of the specimen, but also on its deexcitation mechanisms (fluorescence and phosphorescence) and on its thermal properties; absorption spectra, on the other hand, depend only on the absorption characteristics.

Besides obtaining spectral information on bulk biological matter, one can also use PAS in several ways to investigate the macroscopic structure of the sample. For example, Figure 17.7 shows the PAS spectrum obtained on a piece of apple peel for two chopping frequencies (Rosencwaig, 1978). At the higher frequency, the PAS signal arises solely from the waxy layer at the surface. This layer exhibits mainly a UV absorption due to its protein matter. At the lower frequency, the thermal diffusion length extends below the waxy layer and a PAS signal is obtained from the biological material beneath the waxy layer, such as the carotenoids and chlorophyll compounds in the peel itself. By changing the chopping or modulation frequency in a continuous fashion, one is thus able to perform a depth-profile analysis of biological specimens. Another method for obtaining depth-profile information is to record the PAS spectrum at different phase angles. Kirkbright and his co-workers have performed this experiment on spinach leaf (Adams et al., 1976). They were able to

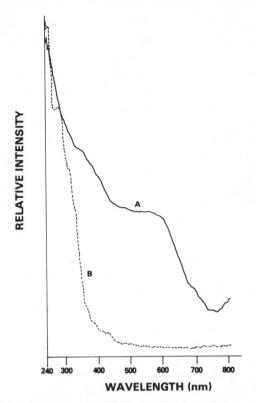

Figure 17.7 Photoacoustic spectra on intact apple peel. The dashed line spectrum was taken at 220 Hz and shows absorption only within the upper waxy layer. The solid line spectrum was taken at 33 Hz and shows absorption within the red peel below the waxy layer as well. (Reproduced by permission from Rosencwaig, 1978.)

demonstrate the existence of the wax layer on the leaf by recording the in-phase and out-of-phase PAS spectra.

Other studies in biological systems are described in Chapter 19, which is concerned with deexcitation studies and in Chapter 20, which discusses thermal properties of materials.

REFERENCES

Adams, M. J., Beadle, B. C., King, A. A., and Kirkbright, G. F. (1976). *Analyst* **101**, 553.

Antonini, E., and Brunori, M. (1971). *Hemoglobin and Myoglobin in Their Reactions with Liquids*, North-Holland, Amsterdam.

Jeffrey, S. W. (1972). *Biochim. Biophys. Acta* **279**, 15.

Jeffrey, S. W., Sielicki, M., and Haxo, F. T. (1975). *J. Phycol.* **11**, 374.

Margoliash, E., and Schijter, A. (1966). *Adv. Protein Chem.* **21**, 113.

Norgard, S., Svec, W. A., Liaaen-Jensen, S., Jensen, A. and Guillard, R. R. L. (1974). *Biochem. Syst. Ecol.* **2**, 7.

Ortner, P. B., and Rosencwaig, A. (1977). *Hydrobiologia* **56**, 3.

Rosencwaig, A. (1973). *Science* **181**, 657.

Rosencwaig, A. (1975). *Anal. Chem.* **47**, 592A.

Rosencwaig, A. (1978). In *Advances in Electronics and Electron Physics*, Vol. 46 (L. Marton, Ed.), Academic Press, New York.

Sauer, K., and Park, R. (1964). *Biochim. Biophys. Acta* **79**, 476.

Scheuer, P. J. (1973). *Chemistry of Marine Natural Products*, Academic Press, New York.

Stewart, W. D. P. (1974). *Algal Physiology and Biochemistry*, University of California Press, Berkeley.

STUDIES IN MEDICINE

18.1 INTRODUCTION

One of the most exciting areas for photoacoustic studies lies in the field of medicine. One can use PAS to obtain optical data on medical specimens that are not amenable to conventional study because of excessive light scattering, to study specimens that are completely opaque to transmitted radiation, and to perform depth-profile analysis by phase or frequency adjustments. Furthermore as we show below, it is possible to construct a photoacoustic cell that can be used to perform all the above measurements in an *in vivo* mode.

18.2 BACTERIAL STUDIES

An example of the use of photoacoustic spectroscopy in medical studies is in the identification of bacterial states. Although conventional light-scattering methods can be used to monitor bacterial growth on various substrates, this very light-scattering makes it most difficult to obtain spectroscopic data on bacteria and thus to identify them. However, bacterial samples are quite suitable for study by PAS. Thus the PAS spectrum of *Bacillus subtilis* var. *Niger,* a common airborne bacteria, contains a strong absorption band at 410 nm when this bacteria is in its spore state (Somoano, 1978). This band is absent when the bacteria is in its vegetative state. Thus photoacoustic spectroscopy enables the detection and discrimination of different bacterial states and allows one to monitor and detect bacteria in various stages of development.

18.3 DRUGS IN TISSUES

Another medical use for PAS is in the study of animal and human tissues, both hard tissues such as teeth and bone, and soft tissues such as skin and muscle.

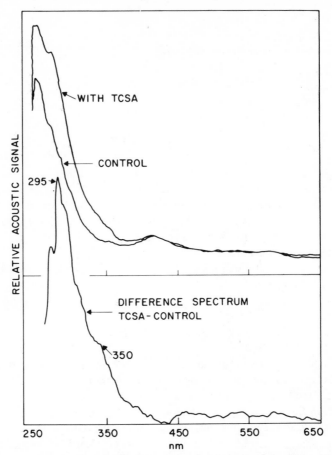

Figure 18.1 A study of guinea pig epidermis. (Upper spectrum) a guinea pig epidermis that has been treated with tetrachlorosalicylanilide (TCSA), (middle spectrum) untreated epidermis, and (bottom spectrum) the difference spectrum showing the absorption bands of the TCSA compound within the epidermis. (Reproduced by permission from Rosencwaig, 1975a.)

An example of a PAS experiment on soft tissue (Rosencwaig, 1975a; 1978) is seen in Figure 18.1, where photoacoustic spectra of guinea pig epidermis are shown. The top spectrum is of an epidermis that has been treated with a 2% solution of tetracholorosalicylanilide or TCSA in ethanol. This compound is known to be a highly effective antibacterial agent, but unfortunately also causes photosensitivity and other skin problems (Jenkins et al., 1964). Why it has these side effects is not completely understood. The central spectrum is of control guinea pig epidermis. The bottom

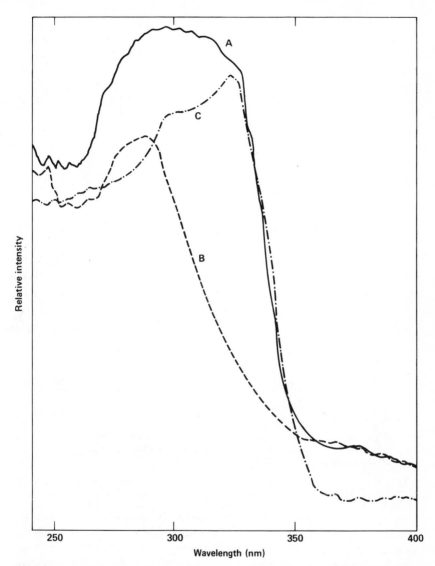

Figure 18.2 A study on sunscreen. (*A*) A PAS spectrum of outer epidermal tissue (stratum corneum) on which a commercial sunscreen has been applied, (*B*) untreated stratum corneum, (*C*) the difference spectrum, giving the spectrum of the sunscreen *in situ*, that is, on and within the epidermal tissue. (Reproduced by permission from Rosencwaig, 1978.)

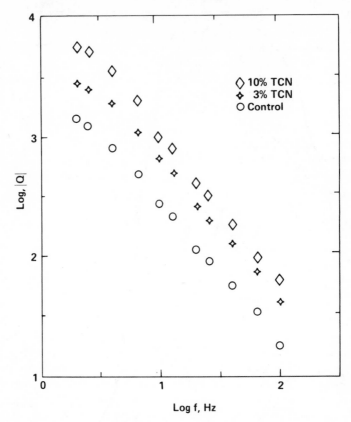

Figure 18.3 Photoacoustic signal dependence on chopping frequency for a control human stratum corneum and for stratum corneum treated with tetracycline (TCN). (Reproduced by permission from Campbell et al., 1977.)

spectrum is the difference spectrum found by subtracting the control spectrum from the TCSA spectrum. The difference spectrum is thus the absorption spectrum of TCSA bound within the epidermis, the first such spectrum obtained by any technique. From this spectrum we can now establish the state of the TCSA compound *in situ*, that is, when it is incorporated into the skin, and thus learn more of its action on and within the skin under various conditions.

Another example of such a study is shown in Figure 18.2 (Rosencwaig, 1978). Here PAS spectra have been taken on human epidermal tissue (stratum corneum) in the 200–400 nm region. The bare stratum corneum is shown in *A*, stratum corneum on which a commercial sunscreen has been applied is shown in *B*, and *C* shows the spectrum of the sunscreen on the stratum corneum as obtained by differential means on a Gilford-R1500 PAS spectrometer (Gilford Instrument Laboratories, Oberlin, Ohio) with a bare stratum corneum sample in the reference cell.

Another example of a drug-in-tissue experiment has been reported by Campbell et al. (1977). In this study the presence of topically applied tetracycline (TCN) has been detected by working at a wavelength of 380 nm, where TCN is fairly strongly absorbing while the skin is not. The experimental results are shown in Figure 18.3, where we see that the PAS signal is substantially greater for the TCN-treated samples than for the untreated sample. The signal amplitude is also roughly proportional to TCN concentration.

These examples illustrate the potential usefulness of photoacoustic spectroscopy in the study of natural, medical, and cosmetic compounds present in and on human tissue.

18.4 HUMAN EYE LENSES

In Figure 18.4 we show PAS spectra taken on intact human eye lenses (Rosencwaig, 1978). These experiments were conducted to study the processes by which human eye cataracts are formed. Little is known about this ancient disease, except that it is probably due to a photooxidative process in which the tryptophan and tyrosine residues in the protein matter of the lens form complexed compounds that are either highly light scattering (cortical cataracts) or colored (brunescent or nuclear cataracts) (van Heyningen, 1975). Cataract studies are severely hampered by the fact that few spectroscopic investigations can be conducted on the intact lenses (Kurzel et al., 1973). In general, these lenses must be solubilized, and except under the strongest solubilizing agents, only one-half of the lens material goes into solution and is suitable for study (Yu and East, 1975). The stronger solubilizing agents do too much damage to the inherent protein to be useful.

The PAS spectra shown in Figure 18.4 were obtained from intact human eye lenses provided by J. Horwitz, Jules Stein Eye Institute, UCLA, and from J. F. R. Kuck, Jr., Emory University Clinic, Atlanta. Both spectra exhibit the characteristic peak at 280 nm due mainly to absorption by the tryptophan and tyrosine protein residues. The cataractous lens shows a much broader 280-nm band than does the normal lens. This is in agreement with the theory that the cataract is a result of conjugated tryptophan and tyrosine compounds (van Heyningen, 1975). In addition, however, the PAS spectra indicate that the cataractous lens also exhibits an increased

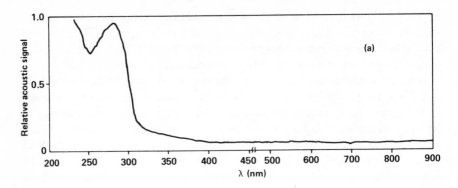

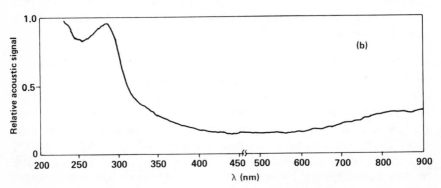

Figure 18.4   A study on intact human eye lenses. (*a*) The PAS spectrum of an intact normal human eye lens. (*b*) The PAS spectrum of an intact human eye lens with a pronounced brunescent cataract. Note that (*b*) indicates that the cataract results in increased optical absorption in both the ultraviolet and infrared regions of the spectrum. (Reproduced by permission from Rosencwaig, 1978.)

infrared absorption. At this time we do not have an explanation for this newly-found feature. It is interesting to note, however, that these spectra indicate that cataractous formation does not only impair vision in the blue end of the spectrum, but actually impairs vision throughout the visible region, with broad absorption bands moving in from both the ultraviolet and the infrared regions of the optical spectrum.

18.5   TISSUE STUDIES

In beginning our photoacoustic studies on human tissues, we hoped that we might be able to use the PAS technique to study abnormal and pathological tissues and thereby obtain new information about diseases such as psoriasis and cancer. However, no such study can be definitive unless adequate baseline data are available on the spectra of normal tissue. Since such data are not available from conventional sources because of light-scattering and opacity effects, it became imperative that we obtain this data using the photoacoustic effect. We therefore began with a study of epidermal tissue, in particular, of the stratum corneum. The experiment below is discussed in detail not only because of the interesting results obtained, but also because of the potential importance of PAS in medical studies on human tissues.

The stratum corneum, the outermost layer of mammalian epidermis, is composed of highly organized units of flat horny cells stacked in vertical interdigitated columns. It is the end result of a specific form of epithelial cell differentiation consisting of synthetic and degradative processes (keratinization) that ultimately form the complex semipermeable and protective matrix consisting of bipolar lipid layers, carbohydrates, proteins, and several other chemical moieties (Matolsky, 1976). The major protein of stratum corneum is α-keratin, which in mammalian epidermis, wool, and other biological systems consists of two or three separate α-helices wound around each other in a coiled-coil configuration (Fraser et al., 1972). Comprehensive treatments of the structure and biochemistry of the stratum corneum have been published recently (Fraser et al., 1972; Mier and Cotton, 1976; Inoue et al., 1976).

The stratum corneum, a translucent membrane in the visible region, is a highly effective light scatterer, especially in the ultraviolet region, and thus cannot be studied readily by conventional optical absorption techniques. Some investigators have attempted to reduce the scattering problem by treating the stratum corneum with fluids of matching refractive index (Lucas, 1931; Runge and Fusaro, 1962). A more popular procedure is to solubilize the sample and then study the resulting optically clear solution. However, this approach is not suitable for studying the stratum corneum in that: (1) the stratum corneum is chemically resistant to complete solubilization because of the strongly cohesive nature of the keratin matrix, and (2) the question of whether the properties of the solution that are measured are exactly the same as those of the unsolubilized sample is the subject of considerable investigation and debate (Rupley, 1969) not only for this system, but for other solubilized systems as well.

To properly characterize the physicochemical properties of the stratum corneum, it is necessary to determine its water content quantitatively. This is a most difficult measurement to perform *in vivo* and most of the successful techniques are performed *in vitro*. The most common method is gravimetric, in which any loss in weight upon heating of the sample is attributed to water. Since the photoacoustic effect is sensitive to the presence of water, through its dependence on the thermal diffusivity of the sample (Chapter 9), we have been able to apply photoacoustic spectroscopy to moisturization (hydration) studies related to the role of water in the stratum corneum (Rosencwaig and Pines, 1977a).

Intact, excised newborn rat stratum corneum has been studied by photoacoustic spectroscopy, as a function of the postpartum age of the rats as well, with rather interesting results (Rosencwaig and Pines, 1977b).

Stratum corneum specimens from newborn rats used in these investigations were obtained 24 hours postpartum for the hydration studies, and from 0 to 60 hours postpartum for the maturation studies. The intact stratum corneum was harvested using Vinson's procedure (Vinson et al., 1964). The excised stratum corneum membranes were typically 1 cm^2 in area and 12–15 microns thick.

For the hydration study, the separated stratum corneum was placed in a dry box for 48 hours, and was subsequently cut into 12 samples, with pairs of samples equilibrated in closed constant humidity chambers of 11, 33, 52, 75, 86, and 93% relative humidity over saturated salt solutions for 24 hours. From each pair, one sample was used for a gravimetric determination of water content (mg H_2O/mg dry stratum corneum), while the other was placed in the photoacoustic cell and its photoacoustic signal was measured at 285 nm.

For the maturation studies, the intact stratum corneum was kept at ambient conditions for at least 2 days and then run in the photoacoustic spectrometer. Several sets of maturation data were taken, each set using the stratum corneum of rats from the same litter, but of different postpartum ages.

Figure 18.5 shows the photoacoustic and optical absorption spectra of a poly-l-glutamic acid membrane in the ultraviolet region. The close similarity seen here between the photoacoustic and the optical absorption spectra was also obtained for the case of collagen membrane and optically clear tape. These results, as well as previous work on biological systems (Rosencwaig, 1973; 1975a; 1975b), clearly indicate the suitability of photoacoustic spectroscopy to obtain optical absorption data on translucent or turbid biological membranes.

In Figure 18.6 we show a typical photoacoustic spectrum, in the 220–450 nm region, of a translucent stratum corneum from a newborn rat. The photoacoustic spectrum in Figure 18.6 is quite clear in spite of excessive light scattering, with very little background signal and no appreciable noise up to ∼220 nm, in contrast to the results obtained with transmission spectroscopy using an integrating sphere (Everett et al., 1966). The so-called protein band in the region of 280 nm is clearly seen. The apparent flatness of the photoacoustic spectrum below 225 nm is due to the saturation effect

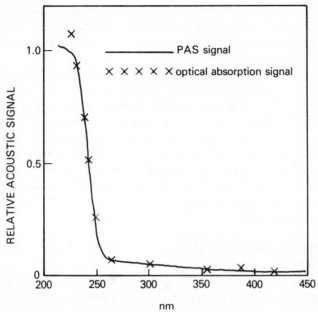

Figure 18.5 Photoacoustic and optical absorption spectra of a poly-1-glutamic acid membrane in the ultraviolet. (Reproduced by permission from Rosencwaig and Pines, 1977b.)

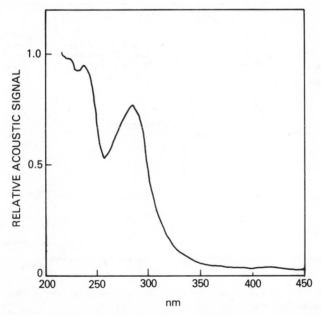

Figure 18.6 Photoacoustic spectrum of newborn rat stratum corneum (~12 μm thick). (Reproduced by permission from Rosencwaig and Pines, 1977b.)

wherein the optical absorption length l_β becomes much smaller than the thermal diffusion length μ (Chapter 9). This occurs because below 225 nm the combined absorption coefficient of all chromophores present becomes very large.

Of the commonly occurring amino acids, only tryptophan, tyrosine, phenylalanine, and cystine possess a characteristic absorption band in the 230–330 nm region. Tryptophan and tyrosine have by far the greatest absorption coefficient in this region. Amino acid analysis on the proteins in the newborn rat stratum corneum gave a consistent value of $\sim$2 tyrosine mole %, and $\sim$0.2 tryptophan mole % using both alkaline and acid hydrolysis. Using these values, and knowing that tryptophan has an absorption coefficient roughly three times greater than tyrosine, we estimate that the 280-nm band seen in Figure 18.6 is $\sim$70% tyrosine, $\sim$20% tryptophan, and $\sim$10% cystine and other indolic and phenolic chromophores.

Figure 18.7 shows that the photoacoustic spectra of solid tryptophan and tyrosine are also quite similar to their solution absorption spectra. The photoacoustic spectra, obtained on powders, exhibit a red shift and a small amount of line broadening relative to the solution spectra, features that are

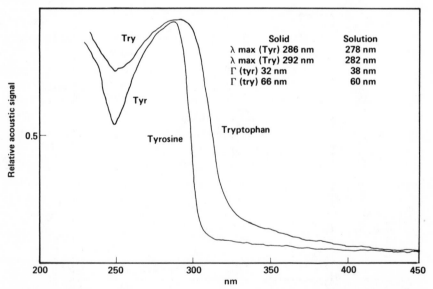

Figure 18.7 Photoacoustic spectra of tryptophan and tyrosine powder. (Reproduced by permission from Rosencwaig and Pines, 1977b.)

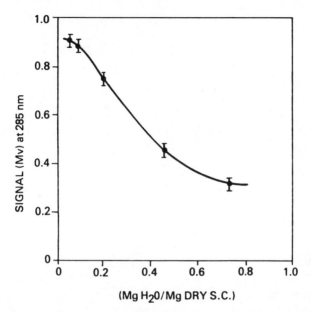

Figure 18.8 Photoacoustic signal strength at 285 nm as a function of water content for newborn rat stratum corneum. (Reproduced by permission from Rosencwaig and Pines, 1977b.)

often seen when the spectrum of a compound in its solid state is compared with that in solution.

Figure 18.8 shows the photoacoustic signal at 285 nm as a function of the water content present in the stratum corneum, as determined gravimetrically. Identical results were obtained at other spectral wavelengths. There is first very little change in the photoacoustic signal, then a fairly rapid decrease, followed by a slower decrease, with the curve apparently approaching a limiting value at high water content.

In Figure 18.9 we show a series of photoacoustic spectra obtained from newborn rat stratum corneum membranes during the initial 60-hour maturation period. The times shown represent the postpartum age of the rats at the time of sacrifice. We note that the 280-nm band undergoes a major change, particularly in the 10–30 hour postpartum period. Identical results were obtained on all sets of rat litters studied, irrespective of the period between the time of harvesting of the stratum corneum and the time the photoacoustic spectra were actually obtained.

Figure 18.10 shows the change in peak position of this 280-nm band as a function of postpartum age, and Figure 18.11 shows the change in band-

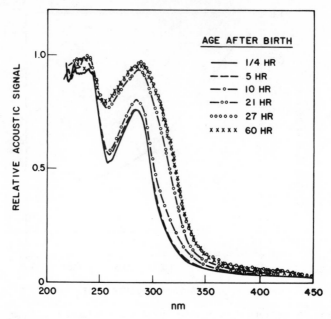

Figure 18.9 Photoacoustic spectra of a series of newborn rat stratum corneum harvested during the postpartum maturation period. (Reproduced by permission from Rosencwaig and Pines, 1977b.)

width and spectral area under this band. Little or no change is observed during the first 10 hours, or after 30 hours postpartum, while a substantial change occurs between 10 and 30 hours postpartum. The band peak begins at ~284 nm, which is close to the value found for tyrosine in a solid matrix, and then increases to 289 nm. The most marked changes are the increases in bandwidth from 38 nm for the $\frac{1}{4}$-hour old rat to a value of ~78 nm for the 30-hour old rat. Normal values are 42 nm for tyrosine in the solid state and 38 nm in solution, with corresponding values of 66 and 60 nm for trypotophan. The area under this band almost trebles, indicating that the band has broadened considerably and that its total absorption or oscillator strength has increased as well.

18.5.1 Hydration Study

The determination of water content in biological tissues would be most useful if it could be performed *in vivo* in a simple and direct fashion.

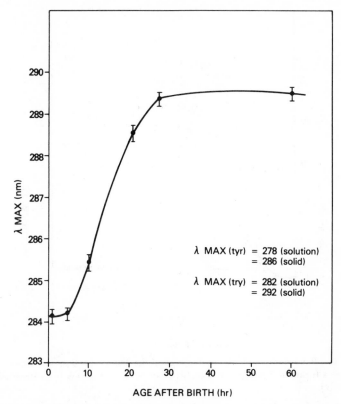

λ MAX (tyr) = 278 (solution)
              = 286 (solid)

λ MAX (try) = 282 (solution)
              = 292 (solid)

AGE AFTER BIRTH (hr)

Figure 18.10 Maximum position of the 280-nm band in newborn rat stratum corneum as a function of postpartum age. (Reproduced by permission from Rosencwaig and Pines, 1977b.)

Photoacoustic spectroscopy is ideal for achieving this purpose, since an open-ended photoacoustic cell can readily be sealed acoustically against a mammalian body, limb, or organ. Furthermore, the dependence of the photoacoustic signal on the thermal properties of the sample makes the technique sensitive to the presence of water.

The shape of the curve shown in Figure 18.8 can be readily understood in terms of the photoacoustic theory of Chapter 9. For a dry sample of stratum corneum, the thermal diffusion length μ is roughly equal to the actual thickness of the sample ($\sim$12 μm). Conditioning the stratum corneum to less than $\sim$60% relative humidity (0.15 mg H_2O) results in only small changes in the cohesive forces maintaining the stratum corneum's rigidity (Papir and Wildnauer, 1974). Moreover, differential

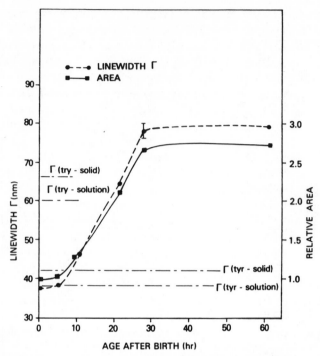

Figure 18.11 Bandwidth and total spectral area of the 280-nm band in newborn rat stratum corneum as a function of postpartum age. (Reproduced by permission from Rosencwaig and Pines, 1977b.)

scanning calorimetry (Walkley, 1972), as well as nuclear magnetic resonance studies, shows that as water is initally added to the stratum corneum, it goes into "bound" sites and, in particular, into the matrix micropores. Since this "bound water" is fairly tightly bound to specific sites, and minimally affects the matrix rigidity (Papir and Wildnauer, 1974), its effect on the total specific heat and thermal conductivity of the stratum corneum is minor in this hydration region. However, its effect on the density of the stratum corneum is not dependent on its binding characteristics, and thus the density increases steadily from the dry state value of ~ 1.0 g/cm^3. In this humidity range, therefore, the net effect of the "bound water" is to decrease the thermal diffusivity ($\alpha = \kappa/\rho C$), thereby decreasing the thermal diffusion length $\mu = (2\alpha/\omega)^{1/2}$.

In Chapter 9 we show that the photoacoustic signal is not noticeably affected by changes in the thermal properties of the stratum corneum until the thermal diffusion length μ becomes smaller than the sample thickness l.

This is apparently the case in our stratum corneum experiment until we reach ~ 0.15 mg HO. When $\mu < l$, the photoacoustic signal is given by (9.36) in Chapter 9, which holds when $\mu < l$ and $\mu \ll l_\beta$. We know from the work of Everett et al. (1966), from the known composition of stratum corneum, and from our own experiments that $l_\beta > l$ in the 280-nm region. Thus the conditions for (9.36) are met, and the photoacoustic signal is given by:

$$q = \frac{-i\beta\mu'}{2} \left(\frac{\gamma P_0 I_0}{2\sqrt{2}l'T_0} \right) \frac{l\mu^2}{\kappa} \qquad (18.1)$$

where μ' is the thermal diffusion length, γ is the specific heat ratio, and P_0 and T_0 are the ambient pressure and temperature, all pertaining to the gas in the cell. I_0 is the light intensity and l' is the gas column length in the cell. Rewriting (18.1) and substituting $\mu^2 = 2\alpha/\omega$ we can set

$$q = \frac{2Z}{\omega} \frac{1}{\rho C} \qquad (18.2)$$

where Z represents all the nonsample parameters and where ρ and C are the density and specific heat of the stratum corneum membrane, respectively. The photoacoustic signal is now independent of the heat conductivity of the membrane, which is known to increase with water content (Bowman et al., 1975).

As we add more water (> 0.15 mg H_2O), the cohesive forces are progressively reduced and the matrix is plasticized, becoming increasingly more extensible and softer (Papir and Wildnauer, 1974). The additional sorbed water behaves more like bulk or "free" water (Walkley, 1972; Bowman et al., 1975). We can thus expect that the specific heat of the hydrated stratum corneum matrix will increase from a value close to 0.3 cal/g-°C (Hove and Kakivaya, 1976) to one close to water itself (1 cal/g-°C). At the same time, the density ρ continues to increase until all the micropores are filled and then begins to level off as it approaches the limiting value of 1.32 g/cm^3 for highly hydrated stratum corneum (Scheuplein, 1967). Therefore, in the hydration region where both ρ and C are increasing fairly rapidly ($0.15 < \text{water} < 0.5$ mg H_2O), the photoacoustic signal decreases fairly rapidly. At even higher humidities (water content > 0.5 mg H_2O), the increases in ρ and C become much more gradual and thus the photoacoustic signal begins to level off as seen in Figure 18.8.

It is apparent from the hydration study that from the use of the theory and some preliminary calibration, one should be readily able to determine water content in biological systems not only *in vitro*, but also *in vivo*.

A similar type of hydration study has been performed by Campbell et al. (1977). In this experiment the thermal constant $(\kappa \rho C)^{1/2}$ was determined as a function of water content at a wavelength and modulation frequency where the sample was photoacoustically opaque, that is, where saturation occurred. Campbell et al. found that the thermal constant $(\kappa \rho C)^{1/2}$ increased with increasing water content, in agreement with the results of Rosencwaig and Pines (1977a).

18.5.2 Maturation Study

Rothman (1954) proposed a theory to the effect that before and/or during the keratinization process of stratum corneum, hydrolysis takes place and some amino acids are incorporated into keratin to form a more highly cross-linked and insoluble keratin matrix, while other amino acids remain free in the cell walls or cellular debris. A preponderance of sulfur-containing amino acids are incorporated into keratin during the keratinization process.

Although most of the published work relating to α-keratin involves wool, the α-keratin of the stratum corneum is quite similar to that in wool (Fraser et al., 1972). Pauling and Corey (1953) have shown that fibrous keratin in wool is predominately α-helical in character, and Crick (1953) has shown that the separate α-helical strands combine to form two-stranded coiled-coil structures in which the two α-helices coil around each other. These coiled-coil filaments are arranged in ring-core tubules, which are held together by a nonfibrous high-sulfur keratin protein matrix. Thousands of these tubules, or microfibrils, are then arranged in a close-packed array, bound together by the keratin protein matrix (Fraser, 1972).

We know how α-helices are formed and stabilized (Pauling and Corey, 1953) and that the strong filament–matrix bonding is primarily due to disulfide bonds (Matolsky, 1976). However, we do not yet know how the coiled-coil formation of the α-helical keratin filaments are stabilized, and as yet there has been no success in making synthetic α-helical polypeptide chains adopt a coiled-coil conformation (Walton and Schodt, 1974).

Mammalian stratum corneum obviously serves quite different roles in the pre- and postpartum periods. Major and rapid biological and structural changes can be expected during the initial postpartum maturation period when the stratum corneum matrix undergoes alteration to develop its so-called barrier functions and adapt to its new and strikingly different environment.

The 280-nm band is primarily due to absorption of the UV radiation by the phenolic (tyrosyl) and indolic (tryptophanyl) chromophores, and in the case of the newborn rat stratum corneum, primarily by the phenolic tyrosine residues. The changes that we see from our photoacoustic spectra in Figure 18.11 therefore reflect changes of the tyrosine, and to some extent tryptophan, residues and changes in their local environment.

Because of excessive light scattering, reliable UV absorption data on intact proteinacious membranes have been, up to now, unavailable. However, there is a voluminous amount of conventional UV absorption data on model compounds and proteins in solution. We make use of this solution data, as well as stratum corneum chemistry, to consider the most likely conventional mechanisms (listed below) to account for our observed spectral changes. We make the reasonable assumption that the initial maturation period is similar to the final stages of keratinization.

1.   An increase in the polarizability of the local environment of the nonpolar amino acids produces a small red shift of the 280-nm band (Wetlaufer, 1962). Such an increase in local polarizability can result from an increase in the sulfur-containing amino acid cystine during the keratinization (maturation) process.

2.   The less-well-developed and somewhat looser keratin matrix structure of the newborn rat becomes more cohesive and tighter as it matures. The indolic and phenolic residues may then find themselves coming closer to negatively charged groups, such as the carboxylate groups. This change in the local environment can alter the absorption characteristics of both tyrosine and tryptophan residues (Wetlaufer, 1962).

3.   Donovan (1973) and Beaven and Holiday (1952) have suggested that several electronic transitions comprise the 280-nm absorption of indole (tryptophanyl) residues, and that one of these transitions may be very sensitive to the local environment. This mechanism is not too promising since our tryptophan content is so low.

4.   Another possible explanation lies in the formation of new hydrogen bonds, such as tyrosyl–peptide carboxyl bonds and tryptophyl–carboxyl bonds (Beaven and Holiday, 1952).

5.   Charge-transfer complexes can often result in both a red shift and some observable line broadening (Donovan, 1973). In most charge-transfer processes in proteins, the "donor" is a side-chain aromatic chromophore, such as tyrosine, and the "acceptor" is a compound containing a substituted benzene ring, a situation that can well be present in a maturing stratum corneum matrix.

Although all the above mechanisms can occur during the initial maturation period, none of them individually, or collectively, appears able to account for the excessive line broadening and the threefold increase in spectral area of the 280-nm band that we have observed. Certainly, major changes in the tyrosine absorption can occur in a highly alkaline solution (Beaven and Holiday, 1952), but not in the pH environment present in biological tissue. In fact, as Donovan (1973) points out, most environmental and protein conformational changes produce such subtle spectral changes of the 280-nm band that difference spectroscopy is often the only means for detecting these changes.

We have also found that our spectral changes in the 10–30 hour postpartum period appear to be correlated to the occurrence of a new protein conformation. This has been indicated by elasticity and differential scanning calorimetry measurements. In the elasticity study it was found that the stratum corneum membranes of rats younger than 10 hours displayed a monotonic change of elasticity with temperature, while the older rats showed an abrupt change in elastic properties at ∼70°C. Similarly, the differential scanning calorimetry showed that the stratum corneum membranes of the older rats (age > 10 hours) displayed a new "protein melting" transition at 70°C, while the younger rats displayed no such transition. It thus appears that the newborn rat stratum corneum undergoes a structural change resulting in a new or modified protein conformation during the 10–30 hour postpartum period.

Since the conventional mechanisms (1–5) appear inadequate to completely account for our results, we have considered some other possibilities.

6. Major spectral changes can be expected to occur if the amino acid composition of the stratum corneum matrix changes substantially during the first 30 postpartum hours. This time period, however, appears much too short for such a change. Moreover, our amino acid analyses showed no noticeable change in tyrosine or tryptophan content during this maturation period.

7. Another possible explanation could be the metabolic synthesis of a strongly absorbing chromophore, such as melanin, during the maturation period. Melanin itself cannot be the chromophore, since we are dealing with albino rats. Furthermore, in Figure 18.12 we show the difference spectrum between a $\frac{1}{4}$-hour and a 30-hour post partum stratum corneum membrane and note that this difference spectrum does not appear to have the spectral features that one would expect from a typical UV-absorbing chromophore. Thus this explanation does not appear too promising, although it cannot be ruled out completely.

8. Another mechanism lies in the possibility that the existing tyrosine residues undergo a major molecular modification, probably by enzymatic

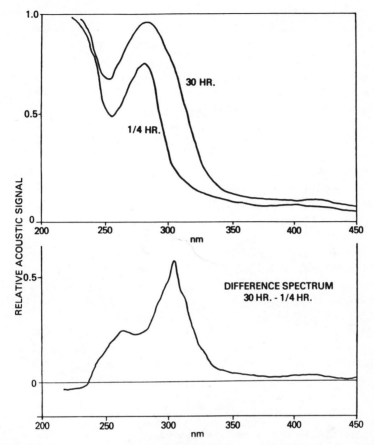

Figure 18.12 (Top) photoacoustic spectra of newborn rat stratum corneum from rats of $\frac{1}{4}$ and 30 hour postpartum age. (Bottom) Difference spectrum between the $\frac{1}{4}$- and 30-hour spectra. (Reproduced by permission from Rosencwaig and Pines, 1977b.)

action. This explanation appears attractive, in that it is well known that a molecular alteration of tyrosine residues within a protein can dramatically change both the shape and the area of the 280-nm band (Gemmill, 1956; Yasunobu et al., 1959). In particular, the action of tyrosinase on tyrosine-containing proteins can alter the 280-nm band in a manner quite similar to that seen in our photoacoustic spectra of Figure 18.9. This is shown very clearly in Figure 3 of Yasunobu et al. (1959), which is reproduced here as Figure 18.13.

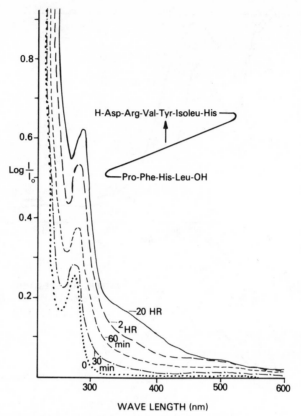

Figure 18.13   The oxidation of hypertensin I by tyrosinase: absorption spectra taken at 0, 30, 60, 120, and 1200 min after addition of the tyrosinase enzyme. (Reproduced by permission from Yasunobu et al., 1959.)

The primary action of tyrosinase on a tyrosine residue bound in a protein is to oxidize it, usually by enzymatically changing the tyrosine to a hydroxyl-tyrosine or DOPA-type molecule. If we hypothesize that an oxidation or hydroxylation of the tyrosine residues occurs in the stratum corneum by some form of enzymatic action (probably by an enzyme other than tyrosinase since we are dealing with albino rats) during the 10–30 hour postpartum period, then not only do we have a mechanism for explaining our photoacoustic data, but we also have a mechanism accounting for the new protein conformational change.

To illustrate this latter possibility, we consider the known amino acid sequence for a single unwound α-helical keratin protein of wool as depicted

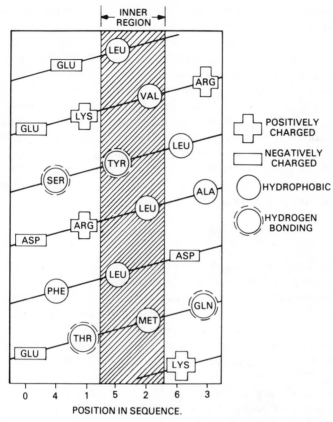

Figure 18.14 Diagrammatic radial projection of the residues in one of the α-helices of a two-stranded coiled-coil keratin filament in wool. (Reproduced by permission from Fraser et al., 1972.)

in Figure 5.4 of Fraser et al. (1972) and shown here in Figure 18.14. The tyrosine residue is located within the nonpolar region of the proposed inner core of the coiled-coil conformation and is shown as being hydrogen bonded within its own helix to a neighboring serine. Thus this tyrosine cannot participate in any interchain bonding.

However, if this tyrosine is now modified to a hydroxyl-tyrosine or DOPA structure, then various interchain hydrogen bonds become immediately available. For example, the hydroxyl-tyrosine could readily hydrogen bond to a lysine or arginine or even another hydroxyl-tyrosine, all on a neighboring helix. Such interchain hydrogen-bonding could play a crucial role in stabilizing the coiled-coil structure of the α-keratin in the stratum corneum.

Although we have as yet no direct evidence for the appearance of a modified tyrosine, such as a hydroxyl-tyrosine, during the maturation or keratinization period, this hypothesis has some attractive features in its analogy with the situation in collagen, the structural protein in muscle tissue. It has recently been determined (Berg et al., 1973; Bansal et al., 1975) that as collagen is being formed, some of the proline residues are enzymatically modified to hydroxyproline, and that the extra hydroxyl group of hydroxyproline contributes significantly to the stability of the triple helix of collagen through additional hydrogen bonds. We feel that the possibility of an enzymatic molecular change in the tyrosine of a postpartum stratum corneum merits serious investigation.

18.5.3 Conclusion

The stratum corneum is a highly complex filament–matrix system with hydrogen bonds, sulfur bonds, salt linkages, electrostatic interactions, and other covalent and noncovalent bonds contributing to its cohesive nature. The study of this complex system is difficult and cannot be pursued by any one method alone. Nevertheless, photoacoustic spectroscopy offers a new and exciting technique for increasing our knowledge of both the fully developed tissue, *in vitro* and *in vivo*, and the maturing tissue. Our new found capability of being able to characterize epidermal tissue noninvasively and *in vivo* and the possibility that we might learn more about the keratinization and postpartum maturation process by the use of photoacoustic spectroscopy and other techniques can play an important role in our understanding of moisturization, substantivity, percutaneous absorption, and normal and diseased states of mammalian tissues.

18.6 COSMETIC APPLICATIONS

PAS has also been applied in mammalian studies to assess the effectiveness of various sunscreens and the substantivity to skin of various formulated sunscreens (Pines, 1978). Again, *in situ* studies by conventional optical means are extremely difficult because of the excessive light-scattering properties of skin in the ultraviolet region.

In his experiment, Pines applied the sunscreen formulations to full-thickness skin that had been excised from rats. After a 30-min post/appli-

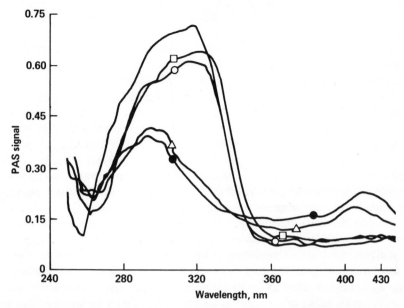

Figure 18.15 Photoacoustic spectra of newborn rat skin. Untreated (●); treated with formulation A, presoaking (−); treated with formulation A, postsoaking (Δ); treated with formulation B, presoaking (□); treated with formulation B, postsoaking (○). (Reproduced by permission from Pines, 1978.)

cation drying time at ambient conditions, a PAS spectrum over the 240–440 nm region was obtained. Immediately following the spectral measurement, the sample was soaked in water and the spectrum was rerun. The results are shown in Figure 18.15.

The presoaking spectra are essentially the same for the two formulations tested. Postsoaking, formulation B retains its initial line shape, whereas formulation A shows a marked change and, in fact, has a spectrum similar to that obtained from the untreated control skin. Clearly, formulation B is more effective since it is not readily removed by water. The effectiveness can be quantitatively determined by the change in spectral intensity with water soaking.

This result is in excellent agreement with clinical sunscreening beach studies and illustrates that one can use photoacoustic spectroscopy to evaluate undiluted topical formulations *in situ* and under "in use" conditions.

18.7   THE *in vivo* CELL

We mention several times in this section the possibility of performing *in vivo* studies with a photoacoustic instrument. The concept for such an instrument is straightforward.

One end of such a cell is open. This end is sealed against the specimen to be studied. This seal may be accomplished either by applying pressure against the specimen or by creating a small negative pressure differential between the outside and the inside of the cell. The optical radiation can be introduced into the cell by means of an optical fiber bundle or by appropriately placed mirrors. With an *in vivo* apparatus, one should operate at chopping frequencies remote from the usual frequencies associated with bodily movements and functions. Nevertheless, it is to be expected that the noise background will probably be higher with this instrument than with the conventional PAS spectrometer. To overcome this difficulty it might prove advisable to increase the incident light intensity or to use more sophisticated data acquisition techniques, such as multiscanning and other averaging procedures. In view of the potential of such *in vivo* studies, these additional procedures should well be worthwhile.

18.8   CONCLUSIONS

Although the use of photoacoustics in the study of mammalian tissues is still in its early stages, this application shows great promise, since tissues are so intractable for study by conventional means. In particular, it may be possible to investigate the spectral properties of abnormal or cancerous tissues and compare these with normal tissues, and from such comparisions obtain new information on the molecular basis of diseased states. Furthermore, one should also keep in mind that the thermal properties of pathological tissues also may be different from those of normal tissues. Since, as we show in Section 5.1 of this chapter, the photoacoustic signal is quite sensitive to changes in the thermal properties of tissues, it may prove possible to use the nonspectroscopic aspects of photoacoustics to investigate abnormal mammalian tissues.

REFERENCES

Bansal, M., Ramakrishnan, C., and Ramachandran, G. N. (1975). *Biopolymers* **14**, 2457.

Beaven, G. H., and Holiday, E. R. (1952). *Adv. Protein Chem.* **7**, 319.

Berg, R. A., Kishida, Y., Kobayaski, Y., Inouye, K., Tonelli, A. E., Sakakibara, S., and Prockop, D. J. (1973). *Biochim. Biophys. Acta* **328**, 553.

Bowman, H. F., Cravalho, E. G., and Wood, M. (1975). *Ann. Rev. Biophys. Bioeng.* **4**, 43.

Campbell, S. D., Yee, S. S., and Afromowitz, M. A. (1977). *J. Bioeng.* **1**, 185.

Crick, F. H. C. (1953). *Acta Crystallogr.* **6**, 685 and 689.

Donovan, J. W. (1973). In *Methods in Enzymology* Vol. 27 (C. A. W. Hirs and S. N. Timasheff, Eds.), Academic Press, New York.

Everett, M. A., Yeargers, E., Sayre, R. M., and Olsen, R. L. (1966). *Photochem. Photobiol.* **5**, 533.

Fraser, R. D. B., MacRae, T. P., and Rogers, G. E. (1972). *Keratins, Their Composition, Structure and Biosynthesis*, Thomas, Springfield, Illinois.

Gemmill, C. L. (1956). *Arch. Biochim. Biophys.* **63**, 177 and 192.

Hove, C. A. G., and Kakivaya, S. R. (1976). *J. Phys. Chem.* **80**, 774.

Inoue, N., Fukayama, K., and Epstein, W. L. (1976). *Biochim. Biophys. Acta* **439**, 95.

Jenkins, F. P., Welta, D., and Barnes, D. (1964). *Nature (Lond.)* **201**, 827.

Kurzel, R. B., Wolbarsht, M. L., and Yamanashi, B. S. (1973). *Exp. Eye Res.* **17**, 65.

Lucas, N. S. (1931). *Biochem. J.* **25**, 57.

Matolsky, A. J. (1976). *J. Invest. Dermatol.* **67**, 20.

Mier, P. D., and Cotton, D. W. K. (1976). *The Molecular Biology of Skin*, Blackwell, Oxford.

Papir, Y., and Wildnauer, R. (1974). *Bull. Am. Phys. Soc.* **19** (2), 264.

Pauling, L., and Corey, R. B. (1953). *Nature (Lond.)* **17**, 59.

Pines, E. (1978). *J. Soc. Cosmetic Chem.* **29**, 559.

Rosencwaig, A. (1973). *Science* **181**, 657.

Rosencwaig, A. (1975a). *Anal. Chem.* **47**, 592A.

Rosencwaig, A. (1975b). *Phys. Today* **28** (9), 23.

Rosencwaig, A. (1978). In *Adv. Electronics and Electron Physics*, Vol. 46 (L. Marton, Ed.), Academic Press, New York.

Rosencwaig, A., and Pines, E. (1977a). *J. Invest. Dermatol.* **69**, 296.

Rosencwaig, A., and Pines, E. (1977b). *Biochim. Biophys. Acta.* **493**, 10.

Rothman, S. (1954). *Physiology and Biochemistry of the Skin*, University of Chicago Press, Chicago.

Runge, W. J., and Fusaro, R. M. (1962). *J. Invest. Dermatol.* **39**, 431.

Rupley, J. A. (1969). In *Structure and Stability of Macromolecules*. (S. N. Timasheff and G. D. Fasman, Eds.), Dekker, New York.

Scheuplein, R. J. (1967). "Molecular Structure and Diffusional Processes Across Intact Epidermis," Final Comprehensive Report No. 7, Directorate of Medical Research, U.S. Army, Chemical Research and Development Laboratory, Edgewater, Maryland.

Somoano, R. B. (1978). *Angew. Chem. Int. Ed. (Engl.)* **17**, 234.

van Heyningen, R. (1975). *Sci. Am.* **233**, 70.

Vinson, L. J., Koehler, W. R., Lehman, M. D., Masurat, T., and Singer, E. J., Eds., (1964). "Basic Studies in Percutaneous Absorption", Semi-Annual Report No. 7, Army Chemical Center, Edgewater, Maryland.

Walkley, K. (1972). *J. Invest. Dermatol.* **59**, 225.

Walton, A. G., and Schodt, K. P. (1974). In *Peptides, Polypeptides and Proteins*, (E. R. Blout, F. A. Bovey, and M. Goodman, Eds.), Wiley, New York.

Wetlaufer, D. P. (1962). *Adv. Protein Chem.* **17**, 303.

Yasunobu, K. T., Peterson, E. W., and Mason, H. S. (1959). *J. Biol. Chem.* **234**, 3291.

Yu, N. T., and East, E. J. (1975). *J. Biol. Chem.* **250**, 2196.

DEEXCITATION PROCESSES IN
CONDENSED MEDIA

19.1 INTRODUCTION

The photoacoustic effect measures the heat-producing deexcitation processes that occur in a system after it has been optically excited, or in more general terms, excited by any electromagnetic radiation. This selective sensitivity of the PAS technique to the heat-producing deexcitation channel can be used to great advantage in the study of fluorescent (or phosphorescent) materials and in the study of photosensitive substances.

19.2 FLUORESCENT STUDIES

When an optically excited energy level decays by means of fluorescence or phosphorescence, then little or no acoustic signal is produced in the photoacoustic cell. This is shown in Figure 19.1, where we consider the case of the fluorescent solid Ho_2O_3 (Rosencwaig, 1975a; 1975b). Several of the trivalent rare earth ions, such as Ho^{3+}, have strongly fluorescent energy levels, that is, levels that tend to deexcite through the emission of a photon, rather than through phonon or heat excitation. The upper PAS spectrum is of some Ho_2O_3 powder containing Co and F impurities. All the lines present in this spectrum correspond to known Ho^{3+} energy levels, whose positions are designated by the bars below. The dots indicate which of these levels are normally fluorescent. In this material the fluorescence is highly quenched by the presence of the Co and F impurities, and thus both the fluorescent and nonfluorescent lines appear in the PAS spectrum. The lower spectrum is of pure Ho_2O_3. Here all the fluorescent levels have a greatly diminished relative intensity, since these levels are now deexciting through the emission of a photon rather than through heating of the solid.

A more illustrative example of the potential of photoacoustic spectroscopy in such studies has been reported by Merkle and Powell (1977), who used photoacoustic spectroscopy to study the radiationless decay processes between the excited states of Eu^{2+} ions in KCl crystals.

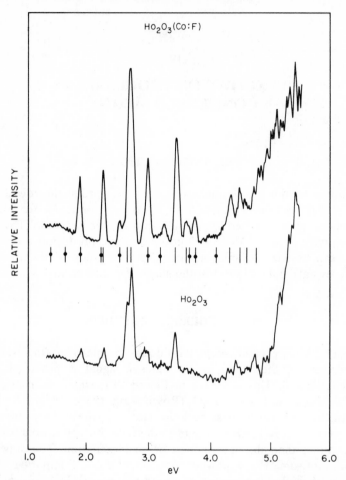

Figure 19.1 Photoacoustic study of a fluorescent material. Cobalt and fluorine impurities quench the natural fluorescence of the Ho^{3+} ions in holmium oxide. (Upper spectrum) doped holmium oxide, and (lower spectrum) undoped holmium oxide. The fluorescent levels are marked with dots. (Reproduced by permission from Rosencwaig, 1975a; 1975b.)

Figure 19.2*a* shows the absorption spectrum of KCl:Eu^{2+} at room temperature. The two strong, broad absorption bands are attributed to transitions from the lowest Stark component of the $^8S_{7/2}(4f^7)$ ground state to the e_g and t_{2g} components of the $4f^65d$ configuration, with the former being at higher energy. The structure on these bands (which is more easily observed at low temperatures) is due to the electrostatic interaction

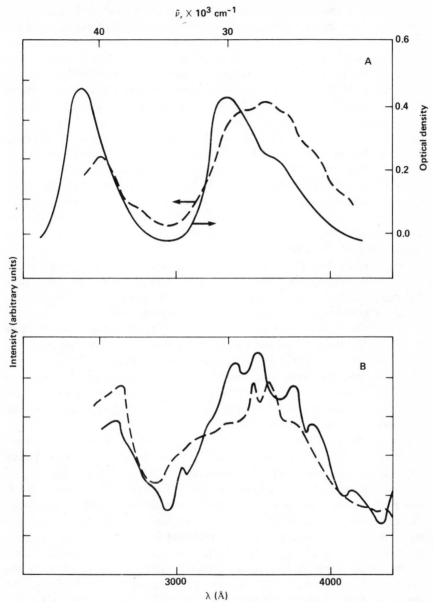

$\tilde{\nu}$, × 10³ cm⁻¹

Figure 19.2 (*a*) Optical absorption (———) and excitation spectra monitored at 4350 Å (_ _ _) for KCl:Eu²⁺ at 300°K. (*b*) Photoacoustic spectra for two different phase shifts: (———) 0° phase, (_ _ _ _ _ _) 45° phase. (Reproduced by permission from Merkle and Powell, 1977.)

247

between the d and f electrons and spin–orbit interaction of the latter. The splitting of the bands indicates the cubic crystal field strength for the d electron is about $10Dq = 12,000$ cm^{-1}, whereas the strength of the electrostatic and spin–orbit effects is found from the fine structure to be on the order of 5000 cm^{-1}. These are dipole-allowed transitions and the radiative decay times for the reciprocal emission transitions from these excited states can be calculated (Fowler and Dexter, 1965) to be on the order of 1 μsec for both the t_{2g} and e_g excited states.

The fluorescence spectrum at room temperature consists of a broad band peaked at approximately 23,800 cm^{-1}, which represents a Stokes shift of $\sim$6000 cm^{-1} from the lowest energy absorption band. The fluorescence decay time at 12°K is found to be about 1.3 μsec and the decay pattern is observed to be purely exponential with no measurable rise time. These results are essentially independent of the wavelength of excitation. The measured fluorescence decay time at low temperatures is close to the predicated radiative decay time.

Figure 19.2a also shows the excitation spectrum at room temperature. Both absorption bands appear, but their relative intensities are quite different. Excitation in the high-energy band leads to a smaller amount of radiative emission that excitation in the low-energy band.

The obvious conclusion to be drawn from these optical spectroscopy results is that the e_g level has two different types of decay channels. The dominant one is total radiationless relaxation to the ground state and the secondary one is a multiphonon transition to the t_{2g} level that fluoresces to the ground state. Both of these processes must take place on a time scale much faster than the $\sim$1 μsec predicated radiative decay time since no radiative emission is detected from this level.

Photoacoustic spectroscopy provides a method for directly monitoring the amount of energy dissipated through radiationless transitions. Figure 19.2b shows the results of PAS measurements made on this system.

Both absorption bands appear in the PAS results. The phase angle at which the signal is maximum is related to the lifetime of the state through the expression $\tau = \tan \phi / 2\pi\nu_c$, where ν_c is the chopping frequency. The signal from the low-energy band is maximum for approximately zero shift in phase from that of the exciting light, which indicates that the decay time of the level is much less than 100 μsec. This is consistent with the measured fluorescence decay time. The peak intensity of the PAS signal in the high-energy band occurs at a phase shift of $\sim$35°, which implies that the lifetime of this state is of the order of milliseconds. It is obvious that this cannot be the radiationless decay time of the e_g level since the radiative decay time is much faster and no radiation is observed from this level.

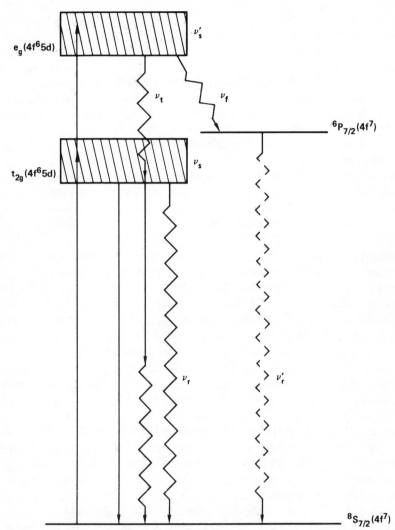

Figure 19.3 Proposed model for excited state relaxation of Eu^{2+} in KCl. (Reproduced by permission from Merkle and Powell, 1977.)

The model energy level diagram shown in Figure 19.3 is proposed to explain the apparent discrepancy between the PAS and optical spectroscopy results. For simplicity the electrostatic and spin–orbit splittings of the $4f^6 5d$ levels are not shown and only one of the manifold of excited $4f^7$ levels is pictured. The photoacoustic signal at phase angle ϕ is simply the sum over all the relaxation transitions that generate heat. For the transition model shown in Figure 19.3, Merkle and Powell estimate that the t_{2g} PAS signal arises from two sources: those electrons undergoing excited relaxation within the t_{2g} band and those undergoing radiationless and vibronic relaxation to the ground state. The e_g PAS signal is somewhat more complicated. It has a contribution due to all the electrons relaxing within the e_g band. Another contribution comes from those electrons that relax to the t_{2g} band and subsequently to the ground state. A third contribution arises from those electrons that relax from the e_g band to the $4f^7$ energy level and then relax radiationlessly to the ground state.

There are thus two excited→ground state radiationless transition times. The one that goes from the t_{2g} to ground (or e_g→t_{2g}→ground) is much faster than 1 μsec, since no radiative emission is detected from this level. The PAS signal that arises from these transitions is at 0° phase shift. The other main relaxation channel is the e_g→$4f^7$→ground state. Absorption, fluorescence, and excitation spectra, along with lifetime data, can be used to predict the relative PAS intensity ratios $I_{e_g}/I_{t_{2g}}$ for different values of the phase angle. This analysis gives a value of the phase angle for the $4f^7$ ground transition of about 66°, in reasonable agreement with experiment. This value implies a $4f^7$ lifetime of 3.6 msec.

The transition model predicted by Merkle and Powell, although rather qualitative, provides an explanation for the apparent discrepancies between optical and photoacoustic data. It thus appears that the dynamics of excited state relaxation in $KCl:Eu^{2+}$ are more complicated than previously thought, and that photoacoustic techniques provide a method for elucidating some of the characteristics of the relaxation processes that cannot be observed by conventional optical means.

In another paper, Peterson and Powell (1978) have reported on a series of PAS studies on various crystals containing Cr^{3+} ions. These crystals included $BaTiO_3$, $SrTiO_3$, MgO, and Al_2O_3. Peterson and Powell compared the radiative and nonradiative modes of decay and found, for example, that the dominant decay mode for Cr^{3+} in MgO is quite different from that in Al_2O_3. They attributed this difference in deexcitation modes to the fact that more than half of the Cr^{3+} ions in MgO are in noncubic sites because of the local defects needed for charge compensation.

19.3 QUANTUM EFFICIENCIES

Since a photoacoustic signal is very sensitive to the presence of a radiative mode of deexcitation, measuring, as it were, the nonradiative complement to the fluorescent or phosphorescent signal, it was soon realized that the PAS technique could be used to measure quantum efficiencies.

The precise determination of absolute fluorescence quantum yields by conventional luminescence means has proven to be very difficult (Calvert and Pitts, 1966). In a luminescence measurement, the number of quanta absorbed from a beam of monochromatic light has to be compared with the number of quanta emitted in the polychromatic fluorescent light, whose distribution in space may be geometrically complicated. This can be accomplished by determining a defined fraction of the fluorescent radiation. To that end various corrections (for geometry, reabsorption, reemission, polarization, refractive index) must be taken into account. This task is tedious and results in low accuracy; errors exceeding 5–10% are common. Details of this method are reviewed by Demas and Crosby (1971).

Another technique involves the measurement of fluorescence lifetime (Dianov et al., 1976). Again, this method suffers from several experimental difficulties, since a separate measurement of the nonradiative contribution to the lifetime of the state must be made, or alternatively the radiative lifetime must be calculated (Riseberg and Weber, 1976).

There is another method that has been somewhat neglected until recently: the determination of the nonradiative part of the absorbed energy by calorimetry (Demas and Crosby, 1971). The temperature rise of an irradiated luminescent sample is compared to the temperature rise of a nonluminescent material showing the same absorption. The main obstacle to this method is the relative insensitivity of common temperature sensors so that strongly absorbing samples usually have to be used. Photoacoustic spectroscopy is another means for performing the calorimetric determination of radiative quantum yield and is generally more sensitive than a conventional calorimetric method.

Lahmann and Ludewig (1977) performed such a measurement on rhodamine 6G in water with a liquid cell containing a piezoelectric type of microphone. In this experiment, the PAS signal from the rhodamine 6G solution was compared with that from a nonfluorescent potassium dichromate solution.

For the nonfluorescent sample, the PAS signal is given by

$$q_1 = Sa_1P_1 \qquad (19.1)$$

where S is the system sensitivity (V/W) and a_1P_1 is the absorbed power, with P_1 the incident power. For the fluorescent solution,

$$q_2 = Sa_2P_2\left[1 - \eta + \frac{\eta(\nu_e - \nu_f)}{\nu_e}\right] \qquad (19.2)$$

where η is the quantum efficiency, ν_e is the optical frequency of the exciting radiation, and ν_f is the mean frequency of the fluorescent radiation. The first part of (19.2) denotes the absorption of nonfluorescent molecules, while the second term gives the heat produced by the fluorescent molecules.

Dividing (19.1) by (19.2) and rearranging yields

$$\eta = \frac{\nu_e}{\nu_f}\left(1 - \frac{a_1 q_2 P_1}{a_2 q_1 P_2}\right) \tag{19.3}$$

A similar experiment was performed by Starobogatov (1977), who used a pulsed ruby laser. The samples studied were also dyes, and the PAS signals from these dyes were then compared with appropriate nonfluorescent solutions. In both the Starobogatov and the Lahmann and Ludewig experiments, the measured quantum yields agreed with those obtained by conventional luminescence experiments, and the accuracy appeared to be better.

Adams et al. (1977) employed a gas-microphone system to obtain the quantum efficiency of quinine bisulfate in aqueous solution. Rather than use another nonfluorescent solution as a standard, they measured the quenching effect of the addition of chloride ions to the sample solution. Again an accurate determination of η was obtained.

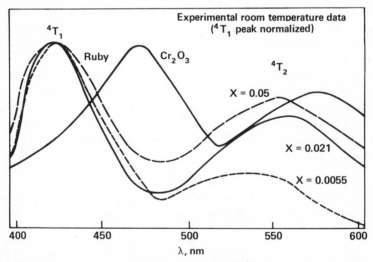

Figure 19.4 Room-temperature PAS spectra for three Cr^{3+} concentrations in ruby and for pure Cr_2O_3. For ease of comparison, the $^4T_1$ peak has been normalized to the Cr_2O_3 spectrum. (Reproduced by permission from Aamodt and Murphy, 1977.)

The first experiment of this nature on a solid sample was reported by Murphy and Aamodt (1977). Here the PAS signal of Cr^{3+} in Al_2O_3 was obtained, and in particular a comparison of the $^4T_1$ and $^4T_2$ band intensities was made as a function of Cr^{3+} concentration as shown in Figure 19.4.

The level diagram for Cr^{3+} appropriate to ruby is shown in Figure 19.5. At low concentrations, light-pumped levels 3 ($^4T_1$) and 4 ($^4T_2$) relax nonradiatively to the single-ion metastable level 2, which relaxes radiatively to ground. With increasing Cr^{3+} ion concentration, some direct nonradiative relaxation occurs from level 4 to ground, and energy is coupled into level 1 as chromium pairs and higher complexes appear. Level 1 also initially relaxes radiatively, but at higher concentrations this emission is quenched through competition with nonradiative relaxation paths.

Using rate equations for the different transitions allowed in Figure 19.5, Murphy and Aamodt were able to relate the ratio of the $^4T_1$ and $^4T_2$ PAS bands to the quantum efficiency η and to obtain data on the quenching effects of increased Cr^{3+} concentration.

Quimby and Yen (1978) also used a gas-microphone method to obtain the quantum efficiency for Nd^{3+} ions in an ED-2 glass matrix. They measured the PAS signal and the fluoroscence lifetime for a number of samples as a function of Nd^{3+} concentrations and from these measurements obtained η for Nd^{3+} ions in the glass sample.

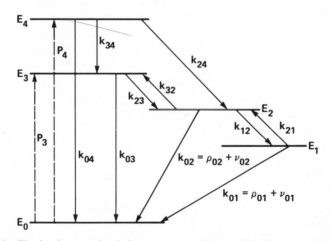

Figure 19.5 Five-level energy-level diagram appropriate to ruby. E_i, p_i, and k_{ij} indicate energies, pumping rates, and total transition rates, respectively; $k_{ij} = \rho_{ij} + \nu_{ij}$, where ρ_{ij} and ν_{ij} are the radiative and nonradiative components of k_{ij}, respectively. (Reproduced by permission from Aamodt and Murphy, 1977.)

The above experiments indicate the type of investigations that can be performed on fluorescent and phosphorescent materials with the photoacoustic technique. A combination of conventional fluorescence spectroscopy and photoacoustic spectroscopy can provide data about both the radiative and nonradiative deexcitation processes within these solids. That is, the complete deexcitation process within these compounds can now be readily studied for the first time. By performing both fluorescence and photoacoustic spectroscopy as a function of temperature and compound composition, one can determine, in a straightforward manner, how these two variables affect the efficiencies and rates for the two deexcitation processes. Furthermore, since photoacoustics gives phase, as well as amplitude, information, one can study exciton processes (random walk, energy level lifetimes, etc.) in these materials as a function of temperature and dopant concentration.

19.4   PHOTOCHEMISTRY

Another channel of deexcitation for absorbed light energy in some compounds is photochemistry. Photoacoustics offers a unique tool for the study of photochemical process in solids. A PAS experiment on the photosensitive material Cooper blue (2,2-dimethyl-4-phenyl-6-p-nitrophenyl-1,3-diazabicyclo(3.1.0)hex-3-ene) (Rosencwaig, 1975a) is shown in Figure 19.6. This compound is colorless in the dark, but turns a strong blue when exposed to light of short wavelength. The bottom spectrum of Figure 19.6 is the PAS spectrum of dark-adapted Cooper blue. There is substantial UV absorption but little visible absorption. The middle spectrum was obtained immediately after the first and is quite different, showing two strong absorption bands in the visible. These are the bands that give Cooper blue its blue color. These bands arise from a photochemical change in Cooper blue wherein some of the photons absorbed in the short wavelength region have been utilized to break a ring in the Cooper blue molecule and thus create a new compound. The upper spectrum, run immediately after the middle spectrum, shows yet further changes, reflecting further photochemical and even photoinduced thermochemical processes.

Not only can one readily see the effects of photochemistry by means of photoacoustic spectroscopy, but one can also establish the activation spectrum for the photochemical process directly, by simply comparing the PAS spectrum with a conventional absorption spectrum. Information about the activation spectrum of photosensitive materials is at present

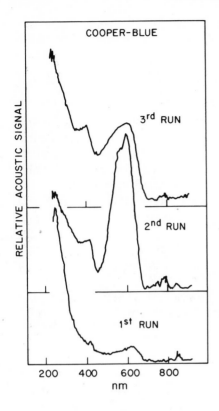

Figure 19.6 Photoacoustic study of a photosensitive material. The Cooper blue was dark adapted before the lower spectrum was run. The middle spectrum was run afterward, followed by the upper spectrum, with only a few minutes separating the different runs. (Reproduced by permission from Rosencwaig, 1975a.)

quite difficult to obtain by other means. In addition, one can obtain, from the phase measurements of the photoacoustic signal, data about photochemical reaction rates, and even distinguish between true photochemical events and photoinduced thermochemical events. Photoacoustic studies on photosensitive materials not only provide valuable basic information about the physical and chemical processes in these materials, but can also be of great benefit in the understanding of technologically important compounds such as photoresists and in the study of photoinduced physical and chemical changes in polymers, plastics, and pigments.

19.5 PHOTOSYNTHESIS

In biology, one of the most important manifestations of photochemistry is the process of photosynthesis both in green plants and in certain bacterial organisms. As in the case of radiative deexcitation, photochemical processes such as photosynthesis compete with the photoacoustic process. The normalized photoacoustic signal q can be written as

$$q = a\left(1 - \sum_i \left(\frac{\eta_i' \Delta E_i}{Nh\nu}\right) - \left(\frac{\eta \nu_f}{\nu_e}\right)\right) \qquad (19.4)$$

Here a is the fraction of light absorbed by the sample, η_i' is the quantum yield for photochemical reactions, ΔE_i is the molar internal energy change per product formation in this reaction, η is the quantum yield for luminescence, ν_f and ν_e are the frequencies of radiated and absorbed light, respectively, N is Avogadro's number, and h is Planck's constant.

If we neglect the role of radiative deexcitation, it is clear that any photochemical process will result in a decrease of the PAS signal with increasing quantum yield for photochemistry. Furthermore, by varying the modulation frequency and analyzing the phase of the PAS signal, information can be obtained on kinetic parameters and on the energy content of intermediates.

Cahen et al. (1978b) have studied photosynthesis in lettuce chloroplast membranes by means of photoacoustics. Since they did not use a piezo-

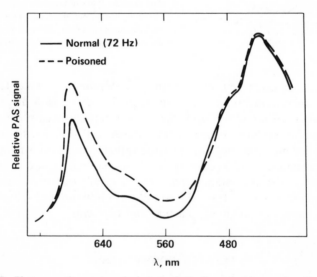

Figure 19.7 Photoacoustic spectra of photosynthetically active (———) and of photosynthetically inactive (_____) lettuce chloroplasts. The photosynthetically inactive samples was immersed in a DCMU-saturated methanol bath and allowed to dry. DCMU is 3-(3,4-dichlorophenyl)-1,-dimethylurea, an electron transport inhibitor. (Reproduced by permission from Cahen et al., 1978b.)

electric cell for the chloroplast suspension, they found that the best results were obtained when the chloroplast suspension was absorbed on cotton wool. Figure 19.7 shows the PAS spectra obtained for photosynthetically active membranes and for DCMU-poisoned [DCMU = 3-(3,4-dichloropheny)-1,1-dimetheylurea, an electron transport inhibitor] chloroplast membranes. The spectra are normalized at 440 nm, where little photosynthetic activity is to be expected. At the 680-nm chlorophyll band, the DCMU-poisoned chloroplasts give a signal that is 10% stronger than that of the active chloroplasts.

As the modulation frequency is decreased, the difference between the spectra of normal and poisoned membranes increases. Cahen et al. explained this observation by hypothesizing that the inhibitory activity of the DCMU does not reach completion at high modulation frequencies.

Another experiment performed by Cahen et al. (1978a) investigated the photosynthetic process in the purple membrane of *Halobacterium halobium*. This membrane contains a single protein, bacteriorhodopsin, covalently bonded to a retinal molecule. Absorption of light by the retinal molecule

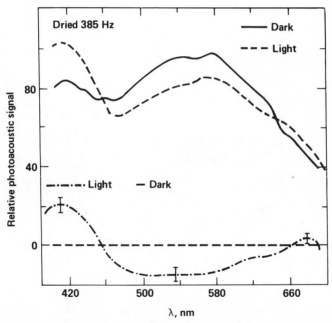

Figure 19.8 PAS of dried purple membrane fragments of *Halobacterium halobium*, with and without 200 mW/cm² continuous side illumination. The bottom spectrum is the difference spectrum (light minus dark). (Reproduced by permission from Cahen et al., 1978a.)

brings about a cyclic photochemical process that drives the translocation
of proteins from one side of the membrane to the other through conformational changes of the protein.

Figure 19.8 shows the PAS spectra of dried purple membrane fragments
with and without strong side illumination. The side illumination causes an
accumulation of the photointermediates absorbing at 415 and 660 nm and
a decrease of the population absorbing at 565 nm.

In lyophillized purple membranes, the absorbed light drives the photocycle only, while in whole cells, parts of the absorbed energy is stored as
ATP and ion gradients. From (19.4) the photoacoustic signal normalized to
the absorbed energy is given by

$$\frac{q}{a} = 1 - \frac{\eta' \Delta E \lambda}{\text{const.}} \tag{19.5}$$

If $\eta' \Delta E = 0$, as would be the case if there were no photochemical reaction
($\eta' = 0$) or if there were a cyclic reaction with $\Delta E = 0$, then q/a is independent of λ. However, if some of the absorbed energy is stored in the

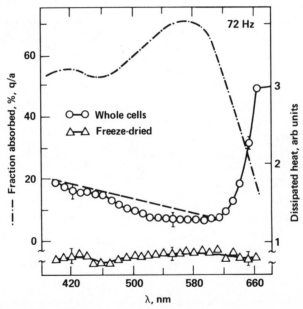

Figure 19.9 Absorption and PAS spectra of whole cells, and PAS spectrum of lyophilized
purple membrane fragments. (Reproduced by permission from Cahen et al., 1978a.)

products sensed by PAS, a valley-shaped curve would be obtained, with the lowest point occurring in the region of highest energy storage.

The results of Cahen et al. appear to demonstrate this situation as seen in Figure 19.9. In the freeze-dried purple membrane fragments, where presumably no energy is stored (on the time scale of the experiment) in the steady state, q/a is constant within experimental error. However, in intact cells q/a passes through a minimum in the 540–620 nm region. These results indicate that photoacoustics appears to be a sensitive tool for the study of photosynthesis. Furthermore, PAS can provide information on the energetics of the process by clearly distinguishing between samples in which part of the absorbed energy is stored or used to do work and those samples in which all the absorbed energy is dissipated in the photocycle as heat.

19.6   NONSPECTROSCOPIC STUDIES OF PHOTOCHEMISTRY

Gray and Bard (1978) have used photoacoustics to study photochemical reactions where the PAS signals were, in the main, attributable to gas evolution or consumption. They studied the oxygen consumption in the photooxidation of rubrene, where singlet oxygen is formed and then attacks the rubrene.

Figure 19.10 shows the diffuse reflectance and the PAS spectrum of 5 wt % rubrene on MgO. If the rubrene from this sample is redissolved in a small quantity of benzene that is allowed to evaporate in air to apparent dryness, enough benzene is retained in the sample to solvate the rubrene and, on exposure to light of wavelengths shorter than 580 nm in the presence of oxygen, the endoperoxidation reaction proceeds. The first scan from long wavelengths in the PAS spectrum of such a sample (Figure 19.10C) shows a large negative-going transient due to oxygen uptake from the gas boundary layer, and this transient is larger than the conventional PAS signal at this wavelength. At shorter wavelengths, the absorbance increases, and the negative O_2-uptake signal is overtaken by the thermal signal. Scans D and F were recorded immediately after scan C. Note that the initial negative-going signal is smaller on the second scan D and has disappeared by scan F. Note also the shoulder that grows at 300 nm, the absorbance peak of the photoproduct. The final scan F looks much like the original scan except for a diminished overall intensity in the rubrene band and the presence of the photoproduct band at 300 nm.

An example of gas evolution was obtained in a study of heterogeneous photocatalytic oxidation of acetic acid to methane and CO_2 at a platinized TiO_2 catalyst. In Figure 19.11, curve A shows the PAS spectrum of the

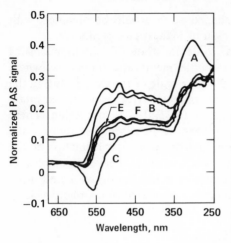

Figure 19.10 PAS of rubrene supported on MgO powder. (*A*) Reflectance spectrum of 5 wt % rubrene in MgO, dried in Roto-vap (ordinate in arbitrary units). (*B*) PAS spectrum of sample in *A*. (*C*) PAS spectrum of sample in *B* solvated with benzene, first scan. (*D–F*) are successive scans following *C*. (Reproduced by permission from Gray and Bard, 1978.)

Pt–TiO$_2$ catalyst in dry powdered form. When the sample is wet with benzonitrile, which does not undergo photodecomposition, the signal level drops because of the presence of the liquid surface layer (curve *B*). However, when the catalyst is wet with acetic acid, the signal level in the region of TiO$_2$ absorption is enhanced considerably as a result of the release of gas from the sample.

These examples demonstrate the utility of PAS in the study of photochemical reactions involving gas evolution and consumption, with a sensitivity of 10^{-13} to 10^{-11} moles of gas evolved or consumed per second.

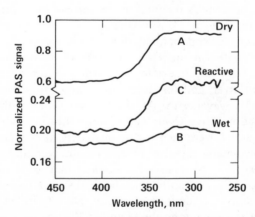

Figure 19.11 Effect of gas evolution on the PAS signals for platinized doped TiO$_2$ (anatase) in the absence and presence of acetic acid. (*A*) Dry TiO$_2$ powder, (*B*)TiO$_2$ powder wet with benzonitrile, (*C*) sample in *B* with 10 μl of acetic acid added. (Reproduced by permission from Gray and Bard, 1978.)

19.7 PHOTOCONDUCTIVITY

In materials where photovoltaic processes can occur, another deexcitation mode competes with the photoacoustic effect (Cahen, 1978). Equation (19.4) can be expanded to include this possibility

$$q = a\left[1 - \sum_i \left(\frac{\eta_i' \Delta E_i}{Nh\nu}\right) - \left(\frac{\eta \nu_f}{\nu_e}\right) - \gamma'\right] \qquad (19.6)$$

where γ' is the energy-conversion efficiency for the photovoltaic process.

In a photovoltaic device, the efficiency γ' is a function of the electrical load on the device. Thus if only the photovoltaic process is present, as for instance in silicon, (19.6) can be written for a specific resistance load R as

$$q(R) = a[1 - \gamma'(R)] \qquad (19.7)$$

where $\gamma'(R)$ can be considered to be the photovoltaic loss. From (19.7) the photovoltaic conversion efficiency is given by

$$\gamma'(R) = \frac{a - q(R)}{a} \qquad (19.8)$$

To solve (19.8) for $\gamma'(R)$ requires a knowledge of a, the energy absorbed by the sample, which is sometimes difficult to obtain. A photovoltaic device, however, has zero energy-conversion efficiency under open-circuit (OC) conditions. Thus $q(oc) = a$, and

$$\gamma'(R) = \frac{q(oc) - q(R)}{q(oc)} \qquad (19.9)$$

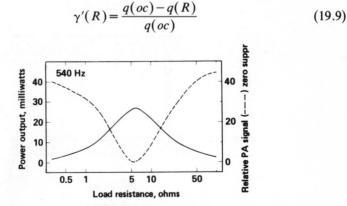

Figure 19.12 Electrical output power (————) and zero-suppressed PAS signal (‒ ‒ ‒ ‒ ‒) as a function of the resistance load on a silicone photovoltaic cell. (Reproduced by permission from Cahen, 1978.)

Figure 19.12 shows Cahen's results on a silicon solar cell, where both the PAS signal and the measured electrical power are shown as a function of the load resistance. As expected from (19.9), the PAS signal is at a minimum when the output electrical power is at a maximum. From this a maximum efficiency of ~17.5% is obtained for the wavelengths used. Finally, Cahen notes that the photovoltaic efficiency can be readily measured as a function of wavelength, as well as load resistance, with photoacoustic spectroscopy.

A nonspectroscopic experiment on photoconductivity has been reported by Ghizoni et al. (1978). In this experiment, the periodic heating of a semiconductor due to electron transport processes in an electric field is studied. Under a d.c. electric field, the free carriers of a semiconductor, gaining energy from the field, ultimately establish a steady state where the energy gained from the field equals the energy lost to the lattice, by means of electron–phonon interactions. Hence, by pulsing a d.c. voltage in a semiconductor, mounted in a PAS cell as shown in Figure 19.13, this periodic heating can be detected by the microphone.

Figure 19.14 shows the acoustic signal as a function of pulse amplitude V_p for two different sample thicknesses and different values of pulse duration τ_p. Figure 19.15 shows the variation of the acoustic signal as a function of pulse duration. These data indicate that at low values of the

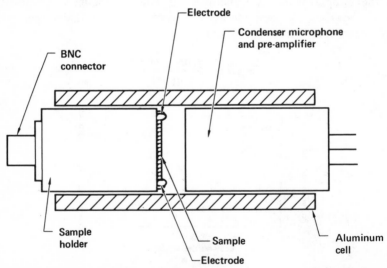

Figure 19.13 Acoustic cell used for investigating the transport properties in semiconductors. (Reproduced by permission from Ghizoni et al., 1978.)

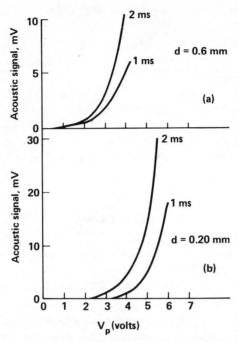

Figure 19.14 Acoustic signal versus the applied peak voltage (V_p) for various pulse durations (τ_p). Sample thickness (a) $d=0.6$ mm, (b) $d=0.2$ mm. (Reproduced by permission from Ghizoni et al., 1978.)

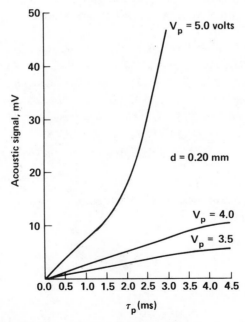

Figure 19.15 Acoustic signal versus pulse duration for different values of the amplitude V_p. (Reproduced by permission from Ghizoni et al., 1978.)

263

electric field, $E = V_p/d$, the acoustic signal is simply given by resistive Jule heating, that is, $q \sim E^2 \tau_p$. However, at high fields, the signal increases exponentially, such that $q \sim e^{E\tau_p}$.

Ghizoni et al. explained these results on the basis that at sufficiently high electric fields, phonon-stimulated emission dominates, that is, the phonon production rate due to electron–phonon collisions becomes larger than the inverse of the phonon relaxation time. Under such conditions, there is phonon gain in the medium, and the photoacoustic signal increases exponentially.

REFERENCES

Adams, M. J., Highfield, J. G., and Kirkbright, G. F. (1977). *Anal. Chem.* **49**, 1850.

Cahen, D. (1978). *Appl. Phys. Lett.* **33**, 810.

Cahen, D., Garty, H., and Caplan, S. R. (1978a). *FEBS Lett.* **91**, 131.

Cahen, D., Malkin, S., and Lerner, E. I. (1978b). *FEBS Lett.* **91**, 339.

Calvert, J. G., and Pitts, J. N., Jr. (1966). *Photochemistry*, Wiley, New York.

Demas, J. N., and Crosby, G. A. (1971). *J. Phys. Chem.* **75**, 991.

Dianov, E. M., Karasik, A. Ya., Neustruev, V. B., Prokhorov, A. M., and Shcherbakov, I. A. (1976). *Sov. Phys. Dokl.* **20**, 622.

Fowler, W. B., and Dexter, D. L. (1965). *J. Chem. Phys.*, **43**, 1768.

Ghizoni, C. C., Sigueira, M. A. A., Vargas, H., and Miranda, L. C. M. (1978). *Appl. Phys. Lett.*, **32**, 554.

Gray, R. C., and Bard, A. J. (1978). *Anal. Chem.* **50**, 1262.

Lahmann, W., and Ludewig, H. J. (1977). *Chem. Phys. Lett.* **45**, 177.

Merkle, L. D., and Powell, R. C. (1977). *Chem. Phys. Lett.* **46**, 303.

Murphy, J. C., and Aamodt, L. C. (1977). *J. Appl. Phys.* **48**, 3502.

Peterson, R. G., and Powell, R. C. (1978). *Chem. Phys. Lett.* **53**, 366.

Quimby, R. S., and Yen, W. M. (1978). *Opt. Lett.* **3**, 181.

Riseberg, L. A., and Weber, M. J. (1976). In *Progress in Optics*, Vol. XV (E. Wolf, Ed.), North-Holland, New York.

Rosencwaig, A. (1975a). *Anal. Chem.* **47**, 592A.

Rosencwaig, A. (1975b). *Phys. Today* **28** (9), 23.

Starobogatov, I. O. (1977). *Opt. Spectrosc.* **42**, 172.

THERMAL PROCESSES

20.1 INTRODUCTION

The photoacoustic process depends not only on the optical properties of the sample, but also on its thermal and geometric properties, and in some cases on its elastic properties as well. Study of the formulae developed in Chapter 9 indicates that the two thermal parameters most commonly encountered in PAS are the thermal diffusivity $\alpha = \kappa\rho/C$ and the thermal constant $D = (\kappa\rho C)^{1/2}$, where κ is the thermal conductivity, ρ is the density, and C is the specific heat.

20.2 THERMAL DIFFUSIVITY

The thermal diffusivity α is of direct importance in heat-flow studies, as it determines the rate of periodic or transient heat propagation through a medium. Because of its controlling effect and common occurrence in heat flow problems, its determination is often necessary, and a knowledge of the thermal diffusivity can in turn be used to calculate the thermal conductivity. An extensive review of thermal parameters has been presented by Touloukian et al. (1973), and some of their recorded thermal diffusivities are presented in Table 20.1.

Adams and Kirkbright (1977) have used the PAS method to obtain thermal diffusivity values for copper and glass by using rear-surface illumination in which a flat-back opaque surface is illuminated and the photoacoustic signal is transferred through a thin copper or glass layer to the gas within a gas-microphone PAS system. The values obtained, 1.68 cm^2/sec for copper and 5×10^{-3} cm^2/sec for glass, compare favorably to those given in the literature.

Changes in the thermal parameters, such as the thermal diffusivity and the thermal constant D, can be used to monitor changes within a material. For example, Rosencwaig and Pines (1977a; 1977b) used the change in thermal diffusivity of rat stratum corneum with water of hydration to obtain data on the role of water in the stratum corneum (see Chapter 18).

TABLE 20.1. Thermal Diffusivities of Condensed Media

Substance	$\alpha(cm^2/sec)$
Liquids	
Alcohol	0.00091
Glycerol	0.000949
Water	0.001465
Solids	
Aluminum	0.982
Aluminum oxide (Al_2O_3)	0.083
Beef	0.00114
Brain tissue	0.00063
Brass	0.337
Brick	0.00278
Carbon (amorphous)	2.06
Glass (plate)	0.0055
Gold	1.30
Iron	0.244
Lead	0.247
Plastic	0.00201
Polyethylene	0.00136
Pyrex	0.00793
Silicon (crystalline)	1.06
Silicon dioxide (SiO_2) powder	0.001952
SiO_2 fused	0.00831
SiO_2 crystalline	0.00894
Silver	1.75
Steel, carbon	0.0639
Steel, stainless	0.0431
Teflon	0.00082

Source: Touloukian et al., 1973.

Campbell et al. (1977) studied the thermal constant D in a similar experiment.

20.3 PHASE TRANSITIONS

Since the thermal parameters of a material generally undergo a change when the material undergoes a phase transition, monitoring the PAS signal as a function of temperature should provide information on the occurence of phase transitions. Florian et al. (1978) were the first to report such a study. They investigated the first-order liquid–solid transition of Ga and of K_2SnCL_6.

Figure 20.1 shows the temperature variation of the PAS signal from gallium (using a nondispersive PAS system) in the temperature region around the melting point ($T_f \sim 29.8°C$). Two important features can be noticed. First, both the amplitude and the phase angle of the PAS signal are different in the liquid and solid phase, and far from T_f they do not depend markedly on the temperature. Second, at T_f an additional strong modification in both amplitude and phase occurs, but only when the first-order transition is approached from lower temperatures.

The first point is simply a consequence of the different values of the thermal parameters in the solid and liquid phases. The second point is explainable from a thermodynamic point of view. During the melting process, all the available heat is absorbed by the melting process, that is, the sample acts as a heat sink for all heat, including the periodic heat pulses generated by the photoacoustic process. Consequently, the PAS amplitude drops markedly and the phase changes drastically. During crystallization, the latent heat released is not detected by the PAS system since it is not at the modulation frequency, but is simply a d.c. signal.

Gallium belongs to those favorable cases where the PAS signal comes from the system that is actually undergoing the phase transition. If the material under investigation is transparent to the incident radiation, phase

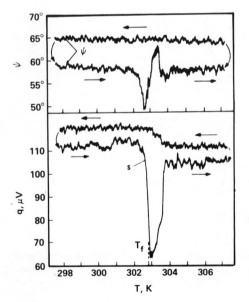

Figure 20.1 Amplitude q and phase angle ψ of the photoacoustic signal of gallium as a function of temperature for both increasing and decreasing temperature [$T_f(Ga) = 29.8°C$]. (Reproduced by permission from Florian et al., 1978.)

transitions can still be examined, although indirectly, through the aid of PAS probes, such as small carbon particles that are in intimate thermal contact with the material under study. The results obtained on water and on K_2SnCl_6 (Figure 20.2) demonstrate the applicability of this indirect method.

For these measurements, carbon black was immersed in water, or mixed with K_2SnCl_6 powder, together with silicon oil. The transition observed in solid K_2SnCl_6 at T_1 is a first-order structural transformation. As compared to those of gallium, the $q(T)$ and $\psi(T)$ curves at the melting point of ice and at T_1 in K_2SnCl_6 are smeared out because of the smaller thermal conductivity of these materials. Although Florian et al. found that the shapes of the q and ψ signals depend on the sample preparation, they were able to demonstrate that the photoacoustic method could define the transition temperature to a fairly high accuracy.

The experiments of Florian et al. demonstrate the ease by which endothermal processes can be detected with photoacoustics. The application of

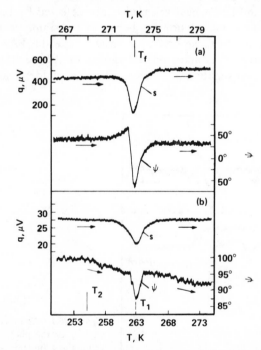

Figure 20.2 Temperature variation (increasing temperature only) of the photoacoustic signal of carbon in (a) H_2O and (b) K_2SnCl_6, at the melting point T_f of ice and at the structural phase transition T_1 of K_2SnCl_6, respectively. (Reproduced by permission from Florian et al., 1978.)

photoacoustics to phase transition studies should constitute a useful complementary technique to the conventional calorimetric methodology.

REFERENCES

Adams, M. J., and Kirkbright, G. F. (1977). *Analyst* **102**, 281.

Campbell, S. D., Yee, S. S., and Afromowitz, M. A. (1977). *J. Bioeng.* **1**, 185.

Florian, R., Pelzl, J., Rosenberg, M., Vargas, H., and Wernhardt, R. (1978). *Phys. Status Solidi* **48**, K35.

Rosencwaig, A., and Pines, E. (1977a). *J. Invest. Dermatol.* **69**, 296.

Rosencwaig, A., and Pines, E. (1977b). *Biochim. Biophys. Acta* **493**, 10.

Touloukian, Y. S., Powell, R. W., Ho, C. Y., and Nicolasu, M. C. (1973). *Thermal Diffusivity*, IFI/Plenum, New York.

CHAPTER

21

DEPTH-PROFILING AND
THICKNESS MEASUREMENTS

21.1 INTRODUCTION

From a study of the gas-microphone theory in Chapter 9, it is clear that one can use the photoacoustic effect to perform various kinds of measurements on the sample. It is obvious that one can utilize a photoacoustic spectrometer to obtain optical absorption data on any and all types of materials. The data may be qualitative when parameters such as the thermal diffusivity or the geometric dimensions of the sample are not known, and fully quantitative when they are known. In addition, one can, by changing the chopping or modulation frequency, obtain a depth-profile analysis of the optical properties of a material. At high chopping frequencies, information about the sample near the surface is obtained, while at low chopping frequencies the data come from deeper within the sample. This is a feature unique to the photoacoustic technique. Another unique capability lies in its ability to obtain optical absorption data on completely opaque materials, provided that one can operate at a chopping frequency high enough so that the thermal diffusion length is smaller than the optical path length. The bulk of the present work in photoacoustic spectroscopy is concerned with those types of experiments done to determine the optical absorption properties of materials. Such experiments have been most fruitful, yielding valuable spectroscopic data on inorganic, organic, and biological systems, data that could not be readily obtained by more conventional techniques.

There are, in addition, two other classes of experiments that are of considerable interest and for which the photoacoustic technique is uniquely suited. It is possible, as we show in Chapter 20, to obtain information about the thermal conductivity of a material, by measuring the thermal diffusion length μ through knowledge of l and β. The thermal conductivity is an important physical parameter that is often very difficult to measure, particularly on powders, amorphous materials, and biological samples. Finally, one can also measure l, through a knowledge of β and μ or by keeping β constant and changing μ in a controlled fashion, or vice versa.

270

Such an experiment can be extremely useful for measuring the thickness of thin films on substrates that possess different optical or thermal properties than the films. Measurements of the thermal and geometric parameters of materials could well become a most important practical application of the photoacoustic effect in industry.

In Chapter 20, we discuss measurements on the thermal parameters of the sample. In this section, we describe how depth-profiling can be used both for spectroscopic purposes and to obtain measurements of layer thicknesses.

21.2 DEPTH-PROFILING

A simple example of depth-profiling is described at the end of Chapter 17. There the PAS spectrum of an apple peel is obtained at two frequencies. At the higher frequency, the PAS signal comes primarily from the UV-absorbing waxy top layer. At the lower frequency, the carotenoids and chlorophyll compounds beneath the waxy layer contribute significantly to the photoacoustic signal.

Most applications where depth-profiling would be an important capability involve layered media or samples whose optical or thermal properties vary with depth. The PAS magnitude and phase from a simple two-layer sample, where the top layer is a thin nonabsorbing layer, can be readily derived from the equations in Chapter 9. Adams and Kirkbright (1977) have shown that for the case when the second layer is itself optically and thermally thin, the signal is given by

$$q \simeq \frac{\beta_1 I_0}{\rho_2 C_2 \omega \mu_2} e^{-x/\mu_2} \left(\cos \omega t - \frac{x}{\mu_2} - \frac{\pi}{4} \right) \tag{21.2}$$

where the subscript 1 denotes a parameter of the bottom layer, while the subscript 2 denotes a parameter of the top (nonabsorbing) layer. From (21.1) we can see that the presence of the nonabsorbing top layer decreases the signal obtained and also introduces a phase lag denoted by

$$\psi = \frac{x}{\mu_2} \tag{21.2}$$

Thus the thickness x of this top layer can be readily determined from the additional phase lag ψ.

Adams and Kirkbright (1977) illustrate an application of this dependence by measuring the thickness of polymer films on copper substrates.

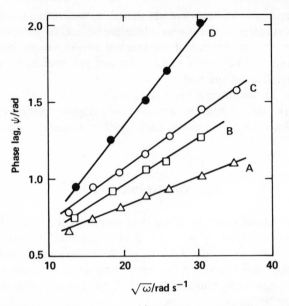

Figure 21.1 Total phase lag ψ_t versus $\omega^{1/2}$ for (A) copper only, and (B, C, and D) copper–polymer films of different polymer thickness. (Reproduced by permission from Adams and Kirkbright, 1977.)

Figure 21.1 shows the total phase lag $\psi_t = x/\mu_2 + F$, where F is a constant, for the copper and copper + polymer samples as a function of $\omega^{1/2}$. The slope of the curves gives $x_t(1/2\alpha)^{1/2}$, where x_t is the total photoacoustic thickness. The polymer thicknesses obtained with a PAS measurement are compared with those obtained by gravimetric (weighting) measurements in Table 21.1. Comparison is reasonable, considering the possible inaccuracies in the α chosen for the thicker polymer films.

TABLE 21.1. Thickness of Polymer Films on Copper

	Film thickness μm	
Sample	By weighing	By PAS
Copper	—	—
Copper + polymer film 1	11 ± 1	9.0 ± 0.6
Copper + polymer film 2	11 ± 1	12.3 ± 1.8
Copper + polymer film 3	30 ± 3	39.0 ± 2.1

Source: Adams and Kirkbright 1977.

There are many situations where the nonabsorbing layer is either very thick or is unavoidably beneath the absorbing layer. In this case, the PAS signal must be analyzed using the general RG equations of Chapter 9. We start by considering the general situation for a homogeneous sample and then consider the case of a two-layer sample.

For these situations, it is imperative to use the exact expressions for the sample temperature developed in the RG theory. To illustrate the use of the exact expressions we present in this section computer-generated plots that give the change in magnitude and phase of the photoacoustic signal as a function of the chopping frequency f, and also as a function of a normalized length L that we define later (Rosencwaig, 1978).

21.3 DEPENDENCE ON MODULATION FREQUENCY

Referring to (9.15), (9.16), and (9.17), and noting that $\kappa' \ll \kappa''$ while a' is usually of the same order of magnitude as a for air at room temperature and pressure, we can ignore g since $g \ll 1$. Also, when the density and specific heat of the backing material are not greatly different than those of the sample, we can set

$$b \simeq \left(\frac{\kappa''}{\kappa} \right)^{1/2} \tag{21.3}$$

Furthermore,

$$a' = a \left(\frac{\alpha}{\alpha'} \right)^{1/2} \tag{21.4}$$

Thus (9.31) can be rewritten as

$$Q = \frac{Z}{a(\beta^2 - \sigma^2)} \left[\frac{(r-1)(b+1)e^{\sigma l} - (r+1)(b-1)e^{-\sigma l} + 2(b-r)e^{-\beta l}}{(b+1)e^{\sigma l} - (b-1)e^{-\sigma l}} \right] \tag{21.5}$$

where

$$Z = \frac{\beta I_0 \gamma P_0}{2\sqrt{2} \, \kappa' l' T_0} \left(\frac{\alpha'}{\alpha} \right)^{1/2} \tag{21.6}$$

and is a frequency-independent term.

It should be noted that the sample-backing combination can also be regarded as a two-layer or thin film–substrate sample, with the bottom layer or substrate not absorbing any of the incident radiation.

In Figures 21.2, 21.3, and 21.4 we show the computer-generated plots derived from (21.5) for the magnitude q and phase ψ. In Figure 21.2 we consider the optical case of a more or less "transparent" sample or first layer, that is, one in which the optical pathlength $l_\beta = 1/\beta \gg l$ (e.g., $l_\beta = 10l$). In Figure 21.3 we have the case of an "absorbing" sample or layer, that is, where $l_\beta \simeq l$, and in Figure 21.4 we have the case of an "opaque" sample or layer, where $l_\beta \ll l$ (e.g., $l_\beta = 0.1l$). In each of Figures 21.2, 21.3, and 21.4 we also consider three cases for the parameter b, which, as defined by (21.3), can be considered as the root of the ratio of the thermal conductivities for the backing or second layer relative to the sample or first layer. The calculations were performed for $b = 0.1$, 1.0, and 10.0, that is, for $\kappa''/\kappa = 10^{-2}$, 1, and 10^2. To generate Figures 21.2, 21.3, and 21.4 we took the values of $l \sim 50$ μm, $\kappa \sim 10^{-3}$ cal/cm-sec-°C, $\rho \sim 2$ g/cm^3, and $C \sim 0.2$ cal/g°C. The values for κ, ρ, and C are typical of reasonably low thermal conductivity solids.

We can see from the log–log plots of the relative magnitude q in Figures 21.2, 21.3, and 21.4 that q varies as f^{-1} for low frequencies and as $f^{-3/2}$ for high frequencies. The change in slope occurs fairly abruptly in the region of 100 Hz for the "transparent" and "absorbing" samples and much more gradually in the region of 1–10 kHz for the "opaque" sample. The effect of b on the frequency dependence of q is minimal, being apparent only at low frequencies where the thermal diffusion length is large enough to encompass the backing or second layer. The effect of the backing or second layer on the frequency dependence of the magnitude of the photoacoustic signal is overshadowed, even at low frequencies, by the stronger frequency dependence of the signal generation process within the sample and at the sample–gas interface.

However, the phase of the photoacoustic signal is, as we can see, quite sensitive to the presence of a boundary in such a two-layer system. We note that there is a phase change of 45° as we proceed from low to high frequencies. The shape of the ψ versus f curve is dependent on both the optical properties of the sample or front layer (β) and the relative thermal conductivities of the two layers (b). The change in phase is fairly abrupt for the "transparent" and "absorbing" samples and occurs in the 1–100 Hz region. It is more gradual for the "opaque" sample and occurs mainly in the 0.1–50 kHz region. We note that for all the optical cases, the ψ versus f curves are sensitive to b only at frequencies $\leqslant 100$ Hz. The explanation for this phenomenon is that only at low frequencies is the thermal diffusion

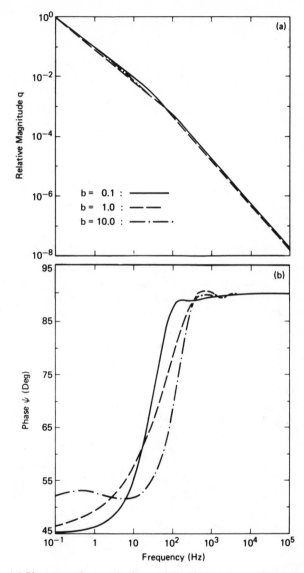

Figure 21.2 (*a*) Photoacoustic magnitude q and (*b*) phase ψ versus chopping frequency f for a "transparent" sample having $l_\beta = 10l$. (Reproduced by permission from Rosencwaig, 1978.)

275

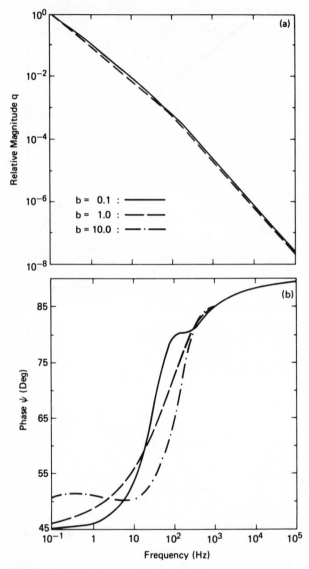

Figure 21.3 (*a*) Photoacoustic magnitude q and (*b*) phase ψ versus chopping frequency f for an "absorbing" sample having $l_\beta \simeq l$. (Reproduced by permission from Rosencwaig, 1978.)

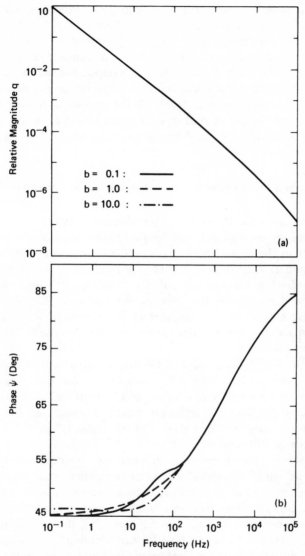

Figure 21.4 (*a*) Photoacoustic magnitude q and (*b*) phase ψ versus chopping frequency f for an "opaque" sample having $l_\beta = 0.1l$. (Reproduced by permission from Rosencwaig, 1978.)

277

length in the sample or first layer large enough to extend into the backing or second layer as well. For the cases of the "transparent" and "absorbing" samples, the major change in ψ occurs at frequencies where the thermal diffusion length becomes comparable to the thickness l of the sample or first layer. For the case of the "opaque" sample, the major change in ψ occurs at frequencies where the thermal diffusion length becomes comparable to the absorption path length of the sample or first layer. Therefore, the presence of the second layer is much less of an influence for the "opaque" sample than for the "transparent" and "absorbing" samples.

21.4   DEPENDENCE ON NORMALIZED LENGTH

To illustrate more clearly how the photoacoustic signal depends on the magnitude of the thermal diffusion length relative to either the geometrical thickness or to the optical pathlength of the sample, we show in Figures 21.5, 21.6, and 21.7 the dependence of q and ψ on the normalized length L. We define $L = l/\mu$ for the case of the "transparent" and "absorbing" samples and $L = l_\beta/\mu$ for the case of the "opaque" sample. We have plotted these curves over six decades of L (equivalent to 12 decades in frequency space) to illustrate that the phase undergoes the 45° change only in the region $L \sim 1$.

As in Figures 21.2, 21.3, and 21.4 the magnitude q varies rapidly with L, with a dependence of L^{-2} for $L < 1$, and of L^{-3} for $L > 1$. The phase is independent of b for the "transparent" and "absorbing" cases when $L > 5$, that is, when the thermal diffusion length is much smaller than the thickness of the sample or first layer. For the "opaque" case, $L = l_\beta/\mu = 0.1$ l/μ and thus is independent of b for $L > 0.5$. In the region of $L \sim 1$, the thermal diffusion length becomes comparable to either the thickness l ("transparent" and "absorbing" cases) or the optical absorption length l_β ("opaque" case), and at this point the phase undergoes a 45° change. The slope of the curve in this region and in the region $L < 1$, is quite sensitive to the thermal properties of the second layer for the "transparent" and "absorbing" cases, but not sensitive for the "opaque" case, since at $L \sim 1$, the thermal diffusion length is already much smaller than l in the "opaque" case.

It is interesting to note that the phase ψ is not a monotonic function of L for the case $b > 1$, that is, for the case when the backing or second layer has a significantly higher thermal conductivity (or $\kappa\rho C$ product) than the sample or first layer.

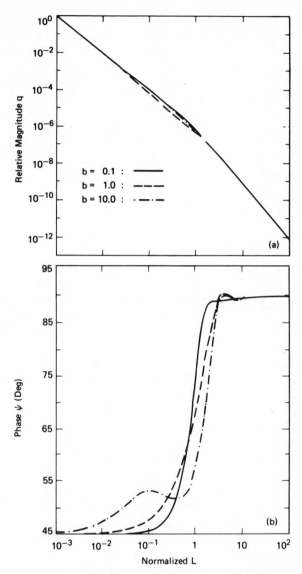

Figure 21.5 (a) Photoacoustic magnitude q and (b) phase ψ versus normalized length $L = l/\mu$ a "transparent" sample having $l_\beta = 10l$. (Reproduced by permission from Rosencwaig, 1978.)

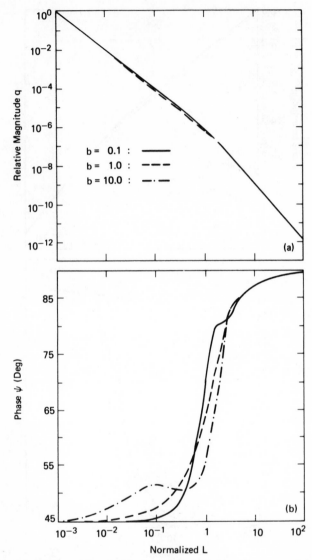

Figure 21.6 (a) Photoacoustic magnitude q and (b) phase ψ versus normalized length $L = l/\mu$ for an "absorbing" sample having $l_\beta \simeq l$. (Reproduced by permission from Rosencwaig, 1978.)

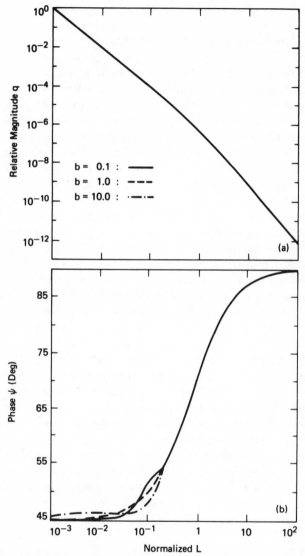

Figure 21.7 (a)Photoacoustic magnitude q and (b) phase ψ versus normalized length $L = l_\beta / \mu$ for an "opaque" sample having $l_\beta = 0.1l$. (Reproduced by permission from Rosencwaig, 1978.)

21.5 NONHOMOGENEOUS SAMPLES

Thus far we have dealt only with the photoacoustic effect in homogeneous solid materials or in layered materials composed of homogeneous layers. There is, however, considerable interest in applying the photoacoustic effect to nonhomogeneous samples as well. Layered systems of interest include thin-film electronic materials, samples coated with paints or polymeric films, and multilayered materials such as photographic film. In addition, there are many problems of practical interest in which the sample is thermally homogeneous and the optical absorption varies continuously with depth from the surface. These problems include the characterization of doped semiconducting material, of laser windows whose surfaces have different absorption properties than the bulk, and of biological tissues.

The analysis of Section 21.4 applies not only to a homogeneous sample in thermal contact with a different backing material, but also to a two-layer system in which the two layers may have both different thermal and different optical properties, and in particular to the case where the lower layer or substrate does not absorb the incident radiation. For example, one can determine the thickness of a film on such a substrate quite accurately if something is known about the film optical absorption coefficient β and its thermal conductivity relative to the substrate (b). Knowing the approximate values of β and b, one can determine the film thickness to within 0.1% for a reasonably strong photoacoustic signal (>0.1 mV at the microphone), if one is working with a "transparent" or "absorbing" film. Furthermore, by measuring the phase as a function of chopping frequency one can, using figures such as 21.2a, 21.3a, and 21.4a estimate the values of b, and from these obtain values for either the thermal conductivity κ or density ρ if the thickness of the film l is known. One can also estimate the optical absorption coefficient β at the wavelength of the incident radiation. The capability of evaluating b from the experimental ψ versus f curves might be of particular interest in monitoring impurity concentrations and structural imperfections in a film, since both of these can alter the thermal conductivity and thus be reflected in the ψ versus f curves through changes in b.

Although the treatment presented in Section 21.4 is a model only for the case of a nonabsorbing substrate (the substrate, however, can be a light scatterer), the RG theory can be formulated as well for the more general case of two or more layers, all of which might absorb the incident radiation, but have different thermal, as well as optical, properties. The unique capability of photoacoustic spectroscopy to perform depth-profile analysis can then be used to full advantage in the study of multilayered and nonhomogeneous substances.

Afromowitz et al. (1977) have considered the case of the thermally homogeneous, but optically nonhomogeneous system. They found that the temperature at the solid–gas interface given in (9.14) can be represented by the expression

$$\theta_0 = \frac{1}{s}\left[\frac{H(s)(b+1)e^{sl}-H(-s)(b-1)e^{-sl}}{(g+1)(b+1)e^{sl}-(g-1)(b-1)e^{-sl}} \right] \qquad (21.7)$$

where $s^2 = i\omega/\alpha$ and $H(s)$ is the single-sided Laplace transfer of $H(x)$, where $H(x)$ is related to the absorption coefficient $\beta(x)$ by the relation

$$\beta(x) = H(x)\left[\frac{(1-R)}{2\kappa} I_0\eta_0 - \int_0^x H(y)\,dy \right]^{-1} \qquad (21.8)$$

where R is the reflection coefficient, I_0 is the incident intensity, and η_0 is the efficiency of the nonradiative transition (usually assumed to be 1). By measuring the photoacoustic signal as a function of chopping frequency ω, one obtains $\theta_0(\omega)$ and thus one obtains $H(s)$ from (21.7). Knowing $H(s)$, one can then invert to obtain $H(x)$, and then using (21.8), one can derive the spatially dependent absorption coefficient $\beta(x)$. Figure 21.8 shows a

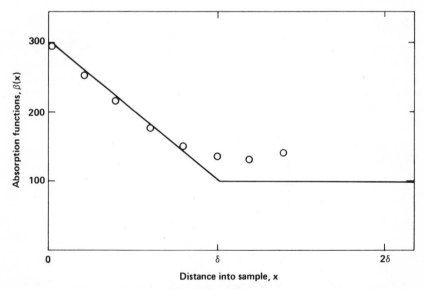

Figure 21.8 Simulated photoacoustic data were calculated at five chopping frequencies for the spatially varying optical-absorption function (solid line). These data were inverted by the procedure described in the text, and the absorption function shown by the points (0) was deduced. (Reproduced by permission from Afromowitz et al., 1977.

test of this treatment. The solid line represents an absorption coefficient that decreases linearly with increasing depth below the surface until the point $x = \delta$, below which the absorption coefficient is a constant value. Simulated photoacoustic data were generated at five chopping frequencies corresponding to thermal diffusion lengths having $\mu = 0.1\delta$, 0.33δ, δ, 3.33δ, and 10δ. The points plotted in Figure 21.8 represent the results of the data-inversion procedure described by Afromowitz et al. (1977). The reasonably good fit to the actual $\beta(x)$ indicates that this method is quite promising for depth-profile analysis of materials by means of the photoacoustic effect.

REFERENCES

Adams, M. J., and Kirkbright, G. F. (1977). *Analyst*, **102**, 678.

Afromowitz, M. A., Yeh, P. S., and Yee, S. S. (1977). *J. Appl. Phys.* **48**, 209.

Rosencwaig, A. (1978). *J. Appl. Phys.* **49**, 2905.

22

EXPERIMENTS AT LOW TEMPERATURES

22.1 INTRODUCTION

Probably because of the newness of the PAS technique and the overwhelming number of applications that are still being explored at room temperature, there has been very little work with photoacoustics at low temperatures. Nevertheless, the few experiments performed at low temperature have been very interesting, and we describe them in this chapter.

22.2 GAS MICROPHONE

Murphy and Aamodt (1977) used a gas-microphone PAS cell at 77°K to obtain data on Cr^{3+} in Al_2O_3. Their low-temperature apparatus is shown in Figure 22.1. The cell consists of a vacuum-insulated double-windowed cylinder to provide optical access to the working volume of the cell. The sound pressure generated within the cell is transmitted to the room-temperature microphone by way of a narrow stainless-steel tube. The entire base of the cell is cooled by direct immersion in liquid nitrogen. Although Murphy and Aamodt reported good results with this system, they did acknowledge that they had problems with acoustic noise from the boiling nitrogen and from inefficient sound propagation in the long tube.

22.3 PIEZOELECTRIC

As yet there have not been any reports of low-temperature experiments using a piezoelectric PAS system. Such a system should offer a considerable advantage over the gas-microphone system, since the piezoelectric detector can operate both at low temperatures and in a vacuum.

22.4 SUPERCONDUCTING BOLOMETER

To overcome the problems of the gas-microphone PAS system at low temperatures and to obtain data on a time scale impossible with micro-

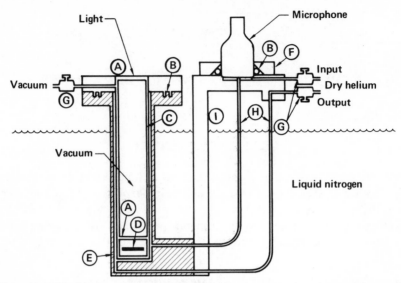

Figure 22.1 Liquid nitrogen PAS cell. (*A*) Quartz windows, (*B*) 0-ring seal, (*C*) thin stainless-steel tubes, (*D*) sample holder, (*E*) sample holder tray, (*F*) microphone lamp, (*G*) vacuum valves, (*H*) needle tubes for sound transfer, (*I*) microphone assembly support. (Reproduced by permission from Murphy and Aamodt, 1977.)

phonics, Robin and Kuebler (1977) devised a heat-pulse technique that utilizes a superconducting bolometer to detect the photoacoustic phonons.

Since the temperature rise of a sample in a typical photoacoustic experiment is of the order of 10^{-6} to 10^{-5} degrees, the most sensitive detector of the direct heat is a superconducting bolometer, a device that undergoes large changes in electrical resistivity for small changes in temperature, provided that the device is at its superconducting transition. Robin and Kuebler used a Pb film $\sim$200 Å thick deposited directly onto the sample. Since the temperature of the Dewar was 3°K, while the T_c for Pb is 7.23°K, a magnetic field was used to tune the superconducting transition down to the Dewar temperature. By illuminating the sample within 0.1 mm of the detector, the response time was between 10 and 100 nsec, limited primarily by the sound velocity in the sample. In actual fact, the experimental response time was of the order of 2 μsec because of the electrical noise emitted by the 3-nsec laser pulse.

In Figure 22.2, Robin and Kuebler show the R_1 and R_2 lines of Cr^{3+} in ruby at 4.2°K as recorded by their PAS bolometer system. From this experiment, they estimate that optical densities as small as 10^{-6} can be detected by this method. Using the heat-pulse technique, Robin and

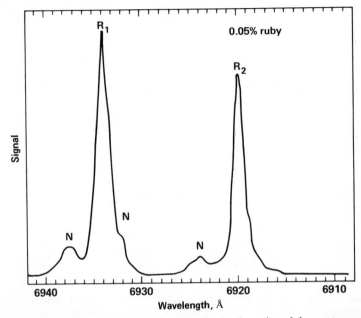

Figure 22.2 Heat-amplitude spectrum of ruby in the R_1–R_2 region of the spectrum at 4.2°K recorded at 38 Hz. The features marked N are due to pair absorption (Reproduced by permission from Robin and Kuebler, 1977.)

Kuebler were also able to obtain lifetime values for several of the energy levels of ruby, as seen in Figure 22.3. As the $^4T_1$ is at a higher energy then the $^4T_2$ line, the very fast pulse in the spectrum for $^4T_1$ is much larger than for $^4T_2$ since more fast deexcitation energy is present. Both the T_1 and T_2 levels deexcite nonradiatively to the ^2E level, which then decays at a much slower rate ($\sim$10 sec) primarily by phosphorescence.

22.5   SEMICONDUCTING BOLOMETER

Parker et al. (1977) decided that the superconducting bolometer as used by Robin and Kuebler was limited by its relatively small dynamic range and the experimental difficulties of maintaining it at its transition temperature. These experimenters found that an amorphous semiconductor film of p-type germanium about 1000 Å thick provided sensitivity comparable to the superconducting bolometer. The detector they made had an intrinsic resistance at 2°K somewhere between the 10 MΩ of a purely amorphous film and the 1 kΩ of a crystalline film. At low temperatures, the resistance

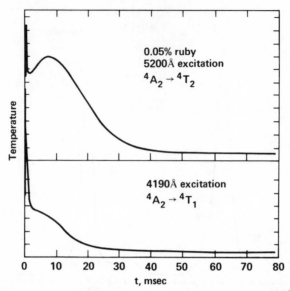

Figure 22.3 Heat-pulse profile in 0.05% ruby stimulated by excitation at 5200 Å (upper) and 4190 Å (lower). (Reproduced by permission from Robin and Kuebler, 1977.)

is a very strong function of temperature. The sample is deposited as a film on a mirrored side of the detector, the mirror preventing any significant heating of the detector by the direct light beam. Parker et al. have measured an NEP (noise equivalent power) for their detector of $\sim 10^{-9}$ W.

REFERENCES

Murphy, J. C., and Aamodt, L. C. (1977). *J. Appl. Phys.* **48**, 3502.

Parker, H., Hipps, K. W., and Francis, A. H. (1977). *Chem. Phys.* **23**, 117.

Robin, M. B., and Kuebler, N. A. (1977). *J. Chem. Phys.* **66**, 169.

PHOTOACOUSTIC MICROSCOPY (PAM)

23.1 INTRODUCTION

The photoacoustic effect, as we state several times earlier in this book, is dependent on the sample's optical characteristics, on its thermal properties, on its geometry, and sometimes on its elastic properties as well. Photoacoustics can thus be used to investigate several different aspects of the same material. It is understandable, therefore, that the possibility of using photoacoustics on a microscopic scale to obtain optical, thermal, geometrical, and elastic images of the sample has occurred to several investigators. Work in this field has only just started, but the results already appear quite promising.

23.2 ULTRASONIC PAM

As we point out in Chapter 10, White wrote a classic paper in 1963 wherein he discussed the generation of high-frequency ultrasonic waves in solid media from the absorption of pulses of electromagnetic radiation. In 1967 Brienza and De Maria demonstrated that Q-switched mode-locked lasers could be used to generate intense ultrasonic waves through surface heating of metal films deposited on piezoelectric crystals. With nanosecond pulse trains they were able to generate sound in solids at frequencies above 2 GHz.

Von Gutfeld and Melcher (1977) were the first to translate White's discovery into a microscopic imaging technique. Their experimental setup is depicted in Figure 23.1. In these experiments, they demonstrated another of White's predictions, that is, the enhancement of the generated acoustic amplitude by the use of mechanically constrained energy-absorbing surfaces. In this work a nitrogen laser with a 10-nsec pulse width or a nitrogen peroxide dye laser with a 5-nsec pulse width was used as the excitation source, with the light focused onto both free and constrained metal surfaces. Constraint could be readily achieved by bonding a transparent dielectric onto the metal surface. Alternatively a thin metallic film

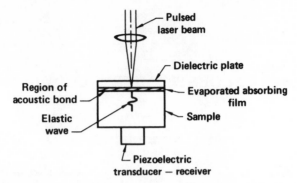

Figure 23.1 Typical structure for generating photoacoustic elastic waves from a constrained boundary using pulsed-laser excitation. (Reproduced by permission from von Gutfeld and Melcher, 1977.)

on a transparent substrate was acoustically bonded to the metal sample with a viscous fluid. Pulse energies of several microjoules were sufficient to produce noticable signals in a 20-MHz piezoelectric detector.

Von Gutfeld and Melcher used these photoacoustically generated elastic waves to detect flaws in a laminate sample. The laminate structure consisted of a top layer about 100 μm thick with no flaws and a bottom layer with holes in it. The detector was bonded to the bottom of the lower layer. Figure 23.2 presents the results obtained. Curves a and c were obtained

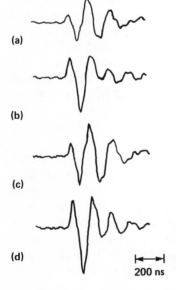

Figure 23.2 Flaw detection of a thin glass disk–Al cylinder laminate. Acoustic patterns are (a.c.) from waves generated by light directed alternately over adjacent holes (0.04-cm diameter) and (b.d.) over spaces between holes. (Reproduced by permission from von Gutfeld and Melcher, 1977.)

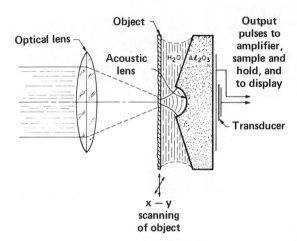

Figure 23.3 Schematic of photoacoustic apparatus. The optical pulse train on the left at 1.06 μm has a peak power of 1 kW. The mode-locked pulses of width $\tau_1 = 200$ psec repeat at 210 MHz ($1/\tau_2$). The Q-switched pulses have a width of 200 nsec with a repetition frequency of 2.7 kHz. The acoustic lens has a radius of 200 μm and the transducer is tuned to 840 MHz. (Reproduced by permission from Wickramasinghe et al., 1978.)

when the laser beam was over the holes in the lower layer, while curves *b* and *d* were obtained between the holes. One can clearly distinguish between these different cases from these curves.

The imaging experiment of von Gutfeld and Melcher, though crude, demonstrated the feasibility of photoacoustic microscopy. In 1978 Wickramasinghe et al. refined the ultrasonic PAM experiment. They modified an acoustic microscope by replacing the input acoustic lens with the beam from a mode-locked, Q-switched Nd:YAG laser, which was focused to a 2 μm spot. The output lens and transducer of the acoustic microscope was then used to detect the photoacoustic signal as shown in Figure 23.3. The transducer was operated at 840 MHz. The sample was rastered mechanically past the beam in two dimensions, and the signal was recorded as a function of position and displayed.

The images shown in Figure 23.4 were obtained from a chromium pattern on a glass cover slip. The chrome was 200 nm thick and was overcoated with 150 nm of aluminum. Thus the object alternated between a layer of aluminum directly on the glass and a layer of chrome overlaid with aluminum. The contrast in the PAM image comes from two factors: (*1*) the photoelastic constant for the aluminum–glass interface is different from that for the chrome–glass interface, and (*2*) the acoustic impedance of the chrome–aluminum double layer is different from that of the single

Metallized
pattern on
glass substrate

(a) Optical (transmission)   (b) Optical (reflection)

(c) Photoacoustic

Figure 23.4 A comparison of the optical and photoacoustic image of a metallized pattern deposited on a glass cover slip. The hexagonal grid pattern is formed with a 200-nm layer of chrome. The cover slip is then overlaid with a 150-nm film of aluminum. The bar width of the hexagonal grid is 25 μm. The quality of the photoacoustic image is degraded through improper synchronization of the CRT and the sample motion. (Reproduced by permission from Wickramasinghe et al., 1978.)

layer of aluminum; as a result, the amount of sound transmitted differs for the two regions. As in the experiment by von Gutfeld and Melcher, the imaging here is the result of the interaction of the photoacoustically-generated ultrasonic waves with the subsurface features.

23.3   GAS-MICROPHONE PAM

The photoacoustic microscopes described above operate at very high ultrasonic frequencies. Wong et al. (1978) showed that one could convert a standard gas-microphone photoacoustic spectrometer into a gas-micro-phone photoacoustic microscope, operating at frequencies below 2 kHz. A schematic diagram of their apparatus is shown in 23.5. With this apparatus, they investigated a number of silicon–nitride ceramic samples, in an attempt to detect the surface cracks that plague ceramic materials. Although the minimum focal point of the laser beam used was only ~30 μm, they were able to detect the presence of surface cracks as seen in Figure 23.6. Even though their experiment was a preliminary investigation of the potentials of PAM, Wong et al. clearly demonstrated the excellent promise of this new technique for nondestructive testing of materials. In this

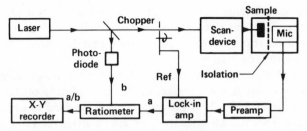

Figure 23.5 Schematic diagram of a laser photoacoustic microscope using a gas-microphone system. (Reproduced by permission from Wong et al., 1978.)

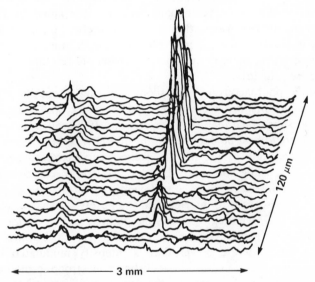

Figure 23.6 An X-Y photoacoustic microscope scan showing the presence of cracks in ceramic surfaces. (Reproduced by permission from Wong et al., 1978.)

experiment, the photoacoustic images were the result of spatial variations in the optical properties of the sample surface. That is, these were essentially optical images.

23.4 PIEZOELECTRIC PAM

To date experiments in photoacoustic microscopy have been performed either with high-frequency ultrasonic systems or with low-frequency gas-microphone systems. The ultrasonic systems tend to be complex, require lasers producing intense pulses of very short duration, and record acoustic images only. The gas-microphone system is limited in frequency and cannot be used readily with large samples, and has been used to obtain optically-determined images only.

Thus far, photoacoustic microscopes have been used to record either an optical image or an acoustic image. However, a photoacoustic microscope operated in the 50 kHz–20 MHz region could be used to obtain thermal-wave images. Thermal-wave images are the result of the interaction of the photoacoustically-generated thermal waves (Figure 9.2) with those features in the sample with different thermal properties. Thus both surface and subsurface features will be imaged by these thermal waves. The ultimate resolution for thermal-wave imaging will be determined by the "wavelength" of the thermal waves, that is, by their thermal diffusion lengths. At 1 MHz, for example, the thermal diffusion length in many materials is of the order of 1 μm, thereby providing microscopic resolution at moderate operating frequencies.

Clearly, a photoacoustic microscope utilizing a simple piezoelectric detector should then provide a novel type of microscopic imaging for both surface and subsurface features (Rosencwaig, 1979). Such a thermal-wave microscope could best be operated in the 50 kHz–20 MHz range. Below 50 kHz, the wavelength of the thermal waves is too large to permit microscopic imaging, while above 20 MHz the acoustic waves, detected by the piezoelectric transducer, have a sufficiently short wavelength such that ultrasonic images begin to interfer with the purely thermal-wave images.

In Figure 23.7 we depict a simple photoacoustic, or thermal-wave, microscope using a piezoelectric detector. The optical beam is a focused laser beam of moderate power that is intensity modulated by an appropriate system, such as an optoacoustic or electrooptic modulator. The sample is rastered by an electromechanical mechanism past the stationary beam. Alternatively, the sample could remain stationary, and the beam could be deflected across the sample by an appropriate deflection mechanism such as an $X–Y$ optoacoustic and an $X–Y$ galvanometric deflection system.

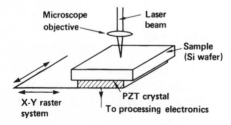

Figure 23.7 A schematic of a simple photoacoustic microscope using a piezoelectric transducer system. (Reproduced by permission from Rosencwaig, 1979.)

The PAM of Figure 23.7 operates with a CW laser and a modulator. In certain cases, it may be preferable to use a pulsed laser with pulse widths in the 50 nsec to 20 μsec range. Also, it should be kept in mind that if a fixed modulation frequency is used, a considerable enhancement in the strength of the piezoelectric signal can be obtained by operating at a resonance frequency of the piezoelectric detector or of the detector–sample combination.

23.5 APPLICATIONS OF PAM IN THE SEMICONDUCTOR INDUSTRY

One of the major uses of a photoacoustic microscope would be in the semiconductor industry, where the instrument could be used to give information about the geometric and material characteristics of Si wafers and their intricate metallization and oxide overlays at various stages of fabrication. Some of these applications are listed below.

1. PAM gives visual information on a microscopic scale. The light is focused to a microscopic spot size. The photoacoustic signal is directly related to the amount of light absorbed at the focused spot. Thus changes in material, or geometric structure, change the absorption or reflection characteristics at the spot and thus alter the photoacoustic signal. A scan of the sample thus gives a picture similar to that obtained with a conventional optical microscope.

2. PAM gives optical absorption data on a microscopic scale. By changing the wavelength of the incident focused light beam, the optical absorption properties of the material at that spot can be analyzed, that is, an optical absorption spectrum can be obtained on a microscopic scale.

3. PAM gives information about local thermal and elastic properties on a microscopic scale. Thus the presence of layered structures, or of faults in metallization or oxide layers, both on and below the surface, are detectable through the associated changes in the local thermal or elastic properties.

4. PAM gives information about deexcitation processes on a microscopic scale. Since the photoacoustic signal arises from the deexcitation of the optical energy levels into localized heat, competing modes of deexcitation affect the PAM signal.

(a) Thus the presence of fluorescent species (e.g., certain dopants or impurities might be ascertained at each microscopic spot, since the presence of fluorescence diminishes the PAM signal. In addition, the fluorescent species might be identified by tuning the wavelength of the incident light through the absorption band (or bands) of the species.

(b) Similarly, the presence of a photovoltaic process, as in the case of semiconductor device materials, also affects the PAM signal. In particular, certain faults in bipolar device manufacture, such as the presence of electrical shorts or leaks, alter the PAM signal and thus become apparent in an early and nondestructive manner. For example, the time dependence of a PAM signal arising from a pulse of light directed on a photovoltaically active region is significantly different if an electrical leak or short is present.

(c) Photochemical processes can be investigated in the same manner as photovoltaic processes.

5. PAM allows for depth-profiling on a microscopic scale. Depth-profiling can be performed in several ways.

(a) By changing the wavelength of the incident light, the depth of optical penetration, and thus the depth at which the photoacoustic signal will be produced, can be changed.

(b) By changing the frequency at which the intensity of the light is modulated, the penetration depth of the thermal waves is altered. This comes about through the dependence of the thermal diffusion length on the modulation frequency. Thus for the case where the optical absorption length is very short ($\beta^{-1} \sim 10^{-6}$ cm), as would be the case for silicon wafers with visible light, the photoacoustic signal can be made to originate from a distance as small as 0.1 μm (10^{-5} cm) at 100 MHz and to as large as 1000 μm (1 mm) at 1 Hz. Full-range depth-profiling thus requires operation over a very broad frequency range.

(c) It is also possible to determine, from a phase analysis of the photoacoustic signal, whether the signal arises from the surface or from the bulk of the material. However, true depth-profiling still requires changing the modulation frequency.

6. An important application of the depth-profiling capability of the photoacoustic microscope is the measurement of thin-film thicknesses on a microscopically localized scale. Such measurements can be performed by analyzing the magnitude and/or phase of the photoacoustic signal as a function of the modulation frequency. Alternatively, these measurements can be performed by analyzing the time dependence of the photoacoustic signal generated by pulses of laser light or pulses from particle beams.

23.6 CONCLUSIONS

There has been some recent work on the possibility of doing photoacoustic microscopy. This work has utilized both the gas-microphone and the piezoelectric methods of signal detection. Probably for most applications the piezoelectric method is to be preferred for photoacoustic microscopy. However, unlike those studies that have used the piezoelectric method only at high ultrasonic frequencies, photoacoustic microscopy will most likely be performed at lower frequencies where thermal-wave microscopic imaging can be done. Finally, there are some important and unique capabilities that a photoacoustic microscope will have if full advantage is taken of the photoacoustic principles.

In particular, photoacoustic microscopy appears to hold considerable promise both as a general analytical tool and as a dedicated process-control instrument for the semiconductor industry. Photoacoustic microscopes can be employed in a semiconductor fabrication line to monitor the presence of electrical shorts or leaks in integrated circuits at a very early stage of the device fabrication. Photoacoustic microscopes can also be used on-line to visualize and inspect patterns and structures that are below the surface and to perform localized thin-film thickness measurements. With these various capabilities, photoacoustic microscopes may well be able to effect a considerable cost savings in the manufacture of large-scale integrated circuits and other electronic devices.

Finally, it should be borne in mind that photoacoustic signals can be generated through the absorption of any and all forms of electromagnetic energy, such as radiofrequency waves, microwaves, infrared, visible, and ultraviolet light, x-rays, and γ-rays. Furthermore, "photoacoustic" signals can also be generated through thermal excitations arising from the interaction with a sample of particle beams, such as beams of electrons, protons, neutrons, ions, atoms, or molecules. Thus other varieties of "photoacoustic" microscopes can be devised that use other forms of electromagnetic radiation, or even particle beams, as the generators of the acoustic signals.

REFERENCES

Brienza, M. J., and DeMaria, A. J. (1967). *Appl. Phys. Lett.* **11**, 44.

Rosencwaig, A. (1979). *Am. Lab.* **11** (4), 39.

von Gutfeld, R. J., and Melcher, R. L. (1977). *Appl. Phys. Lett.* **30**, 257.

White, R. M. (1963). *J. Appl. Phys.* **34**, 3559.

Wickramasinghe, H. K., Bray, R. C., Jipson, V., Quate, C. F., and Salcedo, J. R. (1978). *Appl. Phys. Lett.* **33**, 912.

Wong, Y. H., Thomas, R. L., and Hawkins, G. F. (1978). *Appl. Phys. Lett.* **32**, 538.

FUTURE TRENDS

Photoacoustics and photoacoustic spectroscopy are still in their formative stages. Yet their potential both as research and analytical tools appears almost boundless. Until the development of photoacoustics, many materials, both natural and synthetic, could not be readily investigated by conventional optical methodologies, since these materials occur in the form of powders, or amorphous solids, or as smears, gels, oils, suspensions, and so on. With photoacoustic spectroscopy, optical absorption data on virtually any solid and liquid material can now be obtained.

In this book we review experiments with the photoacoustic effect in the fields of physics, chemistry, biology, and medicine. In all these fields we have done no more than indicate some of the possible applications of this technique to these diverse disciplines.

As photoacoustics becomes better known, investigators in many different fields and with many different problems and orientations will adapt this technique to their own uses. Furthermore, since photoacoustics as a form of spectroscopy is not detector limited, it will surely be extended into the far ultraviolet and infrared regions of the optical spectrum, and quite possibly into other regions of the electromagnetic spectrum as well.

In the near future, it is quite likely that photoacoustics will become a common and useful analytical and research technique in many scientific laboratories, both as a spectroscopic tool and as a nonspectroscopic probe of thermal and elastic properties. Its ease of operation and versatility can only increase its areas of applications. Examples of some new areas are: (*1*) low-temperature studies of organic and inorganic compounds (modification of a PAS cell for low-temperature work should be quite straightforward); (*2*) single-crystal studies, using polarized light, of strongly colored materials that cannot be readily examined by conventional optical techniques; (*3*) transform techniques, for example, the use of Fourier or Hadamard transforms to improve the signal-to-noise ratio when the source intensity becomes weak as in the infrared; (*4*) the field of catalysis, where the characterization of heterogeneous metal oxides and oxide mixtures may prove to be an important application, and where PAS may be employed to monitor ongoing chemical reactions both in the bulk and on the surface of catalysts; (5) the field of biology where the PAS technique can be used to study intact biological systems both in the laboratory and in the field, providing data that now can be obtained only after extensive wet chemical procedures; (6) medicine, where photoacoustics offers the opportunity of extending the exact sciences of noninvasive spectral and thermal analysis to intact medical subtances such as tissues, with the possibility that by such noninvasive techniques new light might be shed on the diseases that afflict mankind.

The next few years promise to be an exciting period of growth for the rediscovered sciences of photoacoustics and photoacoustic spectroscopy.

AUTHOR INDEX

301

SUBJECT INDEX

307